Design Matters

T0214174

James Armstrong

Design Matters

The Organisation and Principles of Engineering Design

 Springer

James Armstrong, OBE, FREng, FICE, FIStructE
32 Langford Green
London
SE5 8BX
UK

ISBN 978-1-84996-595-8   e-ISBN 978-1-84628-722-0

DOI 10.1007/978-1-84628-722-0

British Library Cataloguing in Publication Data
Armstrong, James
Design matters : the organisation and principles
of engineering design
1. Engineering design
I. Title
620'.0042

© 2010 Springer-Verlag London Limited

Apart from any fair dealing for the purposes of research or private study, or criticism or review, as permitted under the Copyright, Designs and Patents Act 1988, this publication may only be reproduced, stored or transmitted, in any form or by any means, with the prior permission in writing of the publishers, or in the case of reprographic reproduction in accordance with the terms of licences issued by the Copyright Licensing Agency. Enquiries concerning reproduction outside those terms should be sent to the publishers.

The use of registered names, trademarks, etc. in this publication does not imply, even in the absence of a specific statement, that such names are exempt from the relevant laws and regulations and therefore free for general use.

The publisher makes no representation, express or implied, with regard to the accuracy of the information contained in this book and cannot accept any legal responsibility or liability for any errors or omissions that may be made.

Printed on acid-free paper

9 8 7 6 5 4 3 2 1

springer.com

Foreword

Following several years as chairman of the Design Matters Group at the UK Royal Academy of Engineering, James Armstrong prepared at the request of the Academy a booklet outlining the basic strategic principles of engineering design, illustrated by four case studies.

James Armstrong has a wide range of experience relating to many different projects and requiring the integrating of many different disciplines. This experience has been used to demonstrate the basic design process and to show how the same general principles can be used to demonstrate the process on such diverse projects. It also suggests how these principles can be used to set out the strategic process for all engineering projects and incorporate the understanding in undergraduate engineering courses at university.

The principles identified cover fundamental strategic issues. They arise from experience of handling projects, involving planning, financing, and political/social issues as well as basic architectural and engineering matters.

Following the publication of the booklet, James Armstrong was invited by the international publishers Springer to expand the presentation of the three principles of need, vision and delivery into a more comprehensive book, of value to the practising engineer and to engineering educators.

In this book the fundamental principles are defined, with suggestions as to the need for and the process of relating the humanities to the sciences of natural law, and demonstrates this with 12 case studies of a variety of design projects from different engineering disciplines. General advice on the establishment of the three principles is given, together with a detailed analysis of one project, and a suggested format for auditing the design process.

Hugh Norie
Chairman Design Matters Group
Royal Academy of Engineering

Contents

Contents

Contributors

Sheri Besford
Building Design Partnership
  (BDP)
16 Brewhouse Yard
Clerkenwell
London
EC1V 4LJ
UK

Harry Carlin-Smith
Managing Director
BSC Consulting Engineers
Rear 37 Macedon Road
Lower Templestowe
VIC 3107
Australia

Peter Fitzgerald
Randox Laboratories
55 Diamond Road
Crumlin
County Antrim
BT29 4QY
Northern Ireland

Anthony Flint
The Old Rectory
Lurgashall
Nr. Petworth
W. Sussex
GU28 9ET
UK

Kelvin Hinson (Ove Arups)
Flat 5
8 Lyndhurst Gardens
London
NW3 5NR
UK

or

Ove Arups & Partners
13 Fitzroy Street
London
W1T 4BQ
UK

Euan Nichol
School of Architectural,
    Civil and Mechanical
    Engineering
Faculty of Health,
    Engineering and Science
Victoria University
Melbourne
Victoria 8001
Australia

Hugh Norie
103 Colehern Court
Old Brompton Road
London
SW5 0ED
UK

Keith Papa
Building Design Partnership
    (BDP)
16 Brewhouse Yard
Clerkenwell
London
EC1V 4LJ
UK

John Roberts
Babtie Group
Fairbairn House
Ashton Lane
Sale
Manchester
M33 6WP
UK

Ralph Waller
Harris Manchester College
University of Oxford
Mansfield Road
Oxford
OX1 3TD
UK

Saaed Zahedi
Chas A Blatchford & Sons Ltd.
Lister Road
Basingstoke
Hampshire
RG22 4AH
UK

1 Introduction

 Purpose and Structure of the Book

The effect of engineering decisions upon the quality of life for the global community is undoubted. It is important therefore that engineers play a full and significant role in ordering the affairs of societies, not only as technicians, carrying out the instructions of others, but as major strategic decision-makers.

The Corporate Plan, 1999–2002, of the Royal Academy of Engineering, the leading UK National Learned Society for engineers, has four key objectives:

"To maintain a crusading, leadership role... recognising the engineers responsibility for leading debate and action on a wide range of important national issues which have an engineering dimension, such as sustainability, biomedical engineering and emerging disciplines..."

"To support engineering education and training at all levels..."

"To work towards greater cross fertilisation between industry and academe..."

"To promote the importance of engineering nationally and internationally; to improve the quality of advice to Parliament and ministries; to promote greater understanding and interest in engineering"

The Royal Charter of the Institution of Civil Engineers in 1818 described the role of the engineer as

"...harnessing the great forces in nature for the use and convenience of Man. "

At the inaugural meeting of the Institution of Civil Engineers, on the 2 January 1818, Henry Palmer said:

"The engineer is a mediator between the philosopher and the working mechanic and like an interpreter between two foreigners, must understand the language of both, hence the absolute necessity of possessing both practical and theoretical knowledge."

To meet the needs and enhance the quality of life of a community, global or local, the engineer, with other complementary creative disciplines, must relate the humanities (the art of understanding the nature of human relationships and societies, of histories and of cultures, of the arts, of planning, of law, of economics) to the natural philosophy of the context, of energy and the elements (earth, water, fire, air and space).

The word "engineer" derives from the word "ingenious", "witty, skilful in invention", "natural capacity", "genius": this developed talent can be deployed in service to the community.

Engineering requires that much time and skill are spent ensuring the delivery of products, projects or services to a required performance and quality specification, on time and within budget. A great deal of the education and training of the engineer is devoted rightly to ensuring his or her ability to effect such a delivery.

But because of this practical focus, the importance of the early decision-making processes is frequently not appreciated, and major decisions are left in the hands of the non-engineering professions, the politicians, lawyers, accountants or marketing experts. The creative and analytical skills of the engineer may be used only to develop, or make practical, the decisions of others.

To meet the objectives set out by the Royal Academy and the Institution of Civil Engineers, there is a need to become more aware of the major contribution that the engineer's abilities can make to the primary decision-making sectors. In this book it is intended to show the crucial nature of the design decision-making process, not only in controlling the delivery process, but in the strategic processes affecting the quality of life in society as a whole, and in their impact upon the world in which we live. The Oxford dictionary definition of design is, among other definitions, "a plan, purpose or intention"; this is the meaning intended in the creative response to meeting an identified need.

The historical context of contemporary design and technology is discussed in this book, showing the development of engineering not as a sudden burst of inspiration with the industrial revolution, but as an increased understanding of the relationships between the natural laws and human needs.

During the last few centuries the learned societies, a legacy of the Age of Enlightenment, have developed and enable us to exchange information and collectively to experiment and to extend our application of this wisdom to meet, even more effectively, the needs of society. This period, together with the growth of the industrial society and of learned institutions (the Royal Society *etc.*) followed from the fundamental understanding and study of natural laws and sciences, from the work of Euclid, Archimedes and Pythagoras, and from the introduction of the enlightened understanding of arithmetic and number in the Vedantic Indian tradition into Western Europe by Fibonacci (1179–1240). This period of over 1000 years led to the contemporary world of high-tech industries and advanced information technology.

The presentation of design as the primary component of engineering is addressed, demonstrating the need to understand the context, process and delivery of engineering projects and services. It is intended to increase awareness of the human qualities as part of the design context, of the delight in achieving a satisfactory and valuable product, and of the understanding and direction of the natural laws and forces in creation.

The considerable importance of the engineer's contribution to the quality of life in the community is not very well understood by society in general, and it is essential that attention is given to publicising the contribution whenever appropriate in the media. We live in an increasingly high-tech age, we could not enjoy our present lifestyles without the continuous technological contribution of the engineer.

The concept of design as a decision-making process giving a creative response to need is demonstrated. This includes the perception of need and context – using the designer's experience, curiosity, alertness, patience, breadth of understanding, and persistence. The nature of the care and service required in design for the client, the user, and the environment is examined in the practical case studies in Chapter 4.

The consideration of the consequences of the decisions – from personal to global – is emphasised as the ethical dimension of the engineer's work. This includes an awareness of, and the need to consider in design, the life cycle of the product, addressing such issues as sustainability, recycling and the real values to the client and user of the completed project or product. The history of the development of each project and the awareness of the need are covered.

In Chapter 2 the design process is explained and analysed, setting out the basic principles of design, and how these may be applied in practice.

The case studies given in Chapter 4 set out the general principles of design used by all experienced designers, but not always consciously articulated. Given a statement of such principles and examples of their use, it can be shown how to design projects in a systematic manner, integrating detailed technical competencies as required, to serve human needs. The importance of a multi-disciplinary team working in effective design development is emphasised.

Chapter 5, the project audit, sets out the parameters for an "audit" of the design processes as a continuing check to ensure that all key factors have been considered.

 # Design Context

Human beings are naturally gregarious creatures. Their development has arisen from their ability and willingness to collaborate with one another, sharing their skills and organising themselves as large groups with common aims – this is evident in great works in many cultures – Indian, Egyptian, Babylonian, Inca, Medieval European. The engineer works in this global society, needing to be aware of and to understand its nature and to respect the economic and cultural differences between developed and developing communities.

As engineers we acquire, as part of the process of professional education and training, a knowledge of the natural laws of this creation, of the great forces in nature, and of the laws governing the interactions between the traditional basic elements – earth, water, fire, air, space – and of the development of our understanding of these laws that has taken place in the last 200 to 300 years. These give us a special knowledge of the opportunities and limitations of the context within which we work. But this understanding must be related to the subtleties of the social context – the humanities are as much a part of the context as are the sciences. We have what Socrates called a "divine curiosity" which motivates us to seek an understanding of the nature and purpose of the creation, and of our own role in this process. This is very evident in the works of Leonardo da Vinci, Isaac Newton, Charles Darwin and many others. As engineers we keep good company!

The relationships between education, training and practice are very important. The development of close relationships between academic institutions and engineering practice are very encouraging. The birth of the modern concept of the universities in the tenth/eleventh centuries was a most important step in the development of skills. Engineers stand as intermediaries between abstract science and practical need. They are concerned not only with understanding and harnessing the forces and materials of nature, but with understanding the structure of society and using their skills to improve the quality of life.

Design is a pervasive activity, a specialised branch of decision-making. An early appreciation of the difference between creative design and logical analysis encourages engineers to participate in the earlier decision-making processes of defining need and envisioning possible ways of meeting that need, in addition to contributing to the essential processes of delivering that vision in reality.

The Needs of Society

All engineering works meet some need, even if the driving force is purely economic, aimed at making a profit for someone, there is a demand for the product, which must be met if the work is to be effective. These needs are very varied, but can be broadly divided into three main areas. In demonstrating the process of design to meet these needs, case studies have been identified which spread broadly across these several areas of need. The three areas suggested are: material, social, and intellectual/spiritual. The general nature of these areas of need and some outstanding specific examples are listed below. Details of each of these projects are given in Chapter 4, illustrating the decision-making processes.

Material needs

- The development of a high-throughput automated analyser, winner of the RAEng McRoberts Award 2003
- The design of an intelligent artificial joint

Social needs

- Design of the Channel Tunnel Terminal at Folkestone, Kent
- The Channel Tunnel Rail Link
- The design of the London Eye
- Lesotho Hydro-electric Project
- Timber Arch Bridge, Melbourne

Intellectual/Spiritual needs

- The design of the Bahá'í Temple, Delhi
- Hampden Gurney Junior School, London
- The Bridge Academy, London
- The design of a residential block for the Harris Manchester College, Oxford
- The design of the Angel of the North statue in Northumberland

 # The Creative Designer

Design is a particular aspect of the generalised human activity of decision-making. Engineering design is a precise, ordered process. Human beings have the gift of curiosity linked with enthusiasm. Every time we make a decision we are using the design process, involving conceptual and lateral thinking. It is of great importance to all engineers to realise this – on the one hand it frees them from the feeling that they are "special", cutting them off from the rest of humanity as "specialists", and on the other it frees them to use their practised decision-making skills at the earlier stages of design.

There is perhaps an analogy with the composer, conductor and performer of music. The composer understands the detailed laws of harmony and counterpoint, but his inspiration comes from his professional talent. He is served well by the conductor and performer, indeed his work cannot be properly appreciated without them, but the contribution of the great composer requires a grounding in principles, as well as creative talent.

How does the creative mind think? Creativity requires an elegance of thought. Whilst an essential aspect of the creative process is the freedom to let the mind roam over many different, sometimes apparently random, concepts, the selection of particularly apt and valuable solutions requires a breadth of view of the need as a whole. To understand the relationships between human concerns and the formulation of design concepts to meet specific needs requires an understanding of natural laws together with a breadth of understanding of the context within which we work. The need is for enthusiasm, creative lateral thinking, breadth of vision, patience and perseverance. The importance of good communication skills and collaborative team working is also an essential component, and is demonstrated in the case studies in Chapter 4.

The role of professional institutions in enabling the free exchange of experience and the peer group assessment of professional ability and competence is an important part of the formation of the creative designer. The senior members of such institutions have experience of the totality of social structures and are able to advise and manage the relationships between these structures and the technological skills available to meet social needs. Their ambitions and commercial skills provide a structure for the integration of highly specialised skills within multidisciplinary teams. As qualified professionals, engineers have a duty of care for their clients and for all affected by their work. As with other professions, this is a particular contribution based on their talents and training and is part of the general duty owed by mankind to the world at large, by we who profess to be human!

The development of an effective visiting professorship scheme by the Royal Academy of Engineering is an excellent example of how experienced engineers can transmit their skills and understanding at the undergraduate level, enabling young engineers to relate their technical skills to the needs of society.

Some ways in which senior engineers can, and do, as visiting professors, contribute to the realisation of the Royal Academy of Engineering's objectives are:

- By promoting and assisting in research and development activities in design methodologies and in design teaching, clearly related to the practice of engineering design.
- By regularly and clearly formulating their personal contribution to design teaching, making best practice available for wider dissemination.
- By stimulating the enthusiasm of undergraduate and young engineers.
- By furthering closer collaboration between academe and industry, not only by their own example, but through other activities, such as staff secondments between the two sectors.
- By clearly and vigorously demonstrating the roles engineers can play in the early stages of design – identifying need and conceptual design – based on their own experience.

The preliminary education of engineers in the technical skills needed can be developed in the early years of professional practice to ensure a smooth flow to becoming fully rounded engineers competent to integrate the important specialist knowledge with the broad social structure. The case studies given in Chapter 4 set out the design process for

7

a variety of projects, and form very good illustrations of the process as part of the undergraduate curriculum. Students can be given the opportunity to work through these or similar cases as part of their design development, or perhaps as specific continuing professional development courses after graduation.

There are many excellent texts available setting out rules and guidelines to assist in the design of particular engineering projects and components. In this book it is intended to show how such subject- or discipline-based procedures are related to the initial formation stages of design, enabling engineers to contribute to the early decision-making processes more fully, using their particular talents and training.

Basic Education

Palmer's statement about engineers standing between natural philosophers and the craftsman is a development of the growth of the industrial society. The basic skills of quiet reflection and clear thinking were developed in the Middle Ages as the traditional seven liberal arts.

John of Salisbury, in a book on logic and reason, *The Metalogicon* (1159), stated:

> "The Liberal Arts... enable the comprehension and understanding of all things... and solve all problems"

> "Reason then examines, with its careful study, those things which have been perceived, and which are to be, or have been, commended to memory's custody. After its scrutiny of their nature, reason pronounces true and accurate judgement concerning each of these."

These studies provide a curriculum which spans the development of the humanities of language, rhetoric and reason (dialectic) – known as the trivium – and links them with the sciences of arithmetic (number), geometry (magnitude), harmony (ordered movement, energy) and astronomy (magnitude in motion, gravity, *etc.*) – the quadrivium.

The engineer has tended to develop considerable skill with the quadrivium subjects, but the development of general communication skills and a broad view of the nature of society has not been so thorough. The liberal arts curriculum was designed, as is implicit in the quotation from John of Salisbury, to teach people how to think, and to understand all aspects of human behaviour. This clearly is of importance to engineers seeking to meet human needs. The relevance of each of the seven subjects of the liberal arts to the engineer is set out below. These qualities can be embodied in undergraduate courses by showing how the subject relates to the social needs to be met, without significantly increasing the amount of information to be taught.

The trivium process includes:

- Grammar/Language: The formulation of concepts and needs
  - The presentation of accurate and unambiguous formulation of project briefs
  - Clarity of language in report writing, *etc.*
- Rhetoric: The communication of ideas
  - Consultation, presentation of evidence to planning and parliamentary committees
  - Learned society committee work
  - Teaching, report writing, *etc.*
- Dialectic: The discovery of real needs
  - Exploration of the world of ideas and creative thinking
  - Improvements in "how to think", basic philosophical concepts and methods

The quadrivium subjects relate to natural laws covering:

- Arithmetic: Number at rest
  - Numeracy and logic, the nature of proof
  - Basic arithmetic and algebra
- Geometry: Magnitude at rest
  - Form, proportion, the divisions of space
  - Platonic solids, Euclidean demonstrations, model making
- Music: Number in motion
  - Harmony, energy, power, resonance
  - Thermodynamics, wave motion
- Astronomy: Magnitude in motion
  - Movement, gravity, stress
  - Growth, spiral developments, changes of direction, spatial relationships

These subjects were developed as a basis for general education. To act as a professional agent for the Creator requires that we receive a full education in the humanities, as well as specialist education in particular subjects, such as engineering. These subjects provide for the formulation of need, the creative process of analysis, synthesis, specification, sketches, engineering drawings, surveys, *etc.*, and the delivery of finished designs as physical changes that raise the quality of life in societies. The importance of a balanced education and training of engineers in all these general skills as well as in the technological sciences and crafts is evident in effective creative design.

2 The Design Process

Introduction

There are strategic stages in the design process used by all experienced designers, but not always consciously articulated, which if well understood, certainly express the challenge and delight in engineering. Given a statement of such principles and examples of their use, it is of value to show how to design projects in a systematic manner, integrating detailed technical competencies as particular disciplinary postulates, as distinct from fundamental axioms or principles. In this chapter this systemic process is set out and illustrated.

There can be said to be a hierarchy of principles: from the general to the particular. In this book we are setting out the general strategic principles concerned with the process of design. There is a need for collaborative team work, with "brain-storming" interdisciplinary design sessions. The managerial skills and resources to assemble and control large numbers of people over periods of time are essential components of effective design practice.

Senior engineers are actively engaged in primary decision-making activities. They encourage freedom of thought in the context of structured programmes. They understand the nature of this stage in the meeting of human needs, and have developed the ability to envision solutions to defined needs, and then to ensure the delivery of these solutions.

The abilities of the engineer can make a valuable contribution to primary decision-making. Primary decision-making is the first step in design. It begins with the fundamental decisions as to the need to be met. As engineers it is important that we recognise that this stage benefits from, and indeed requires, a professional engineering contribution. Paradoxically, because of our great success in the delivery process, we have reduced our contribution to primary decision-making by associating the design activity with the development of decisions made by others. This book attempts to increase the awareness of the engineer as to the primary stages of design in meeting society's needs.

During my early years as an engineer, a small group of my contemporaries met regularly each week at lunchtime to present our work for the benefit of each other. This professional enthusiasm continued, and the members of the group remained in touch with one another for over 50 years. This interest in our work is not untypical of young professionals and contributes to the transfer and development of skills and experience.

Design is the essential creative process of engineering, which distinguishes it from science, and which calls for imagination and creativity, as well as the knowledge and application of technical and scientific skills and the skilful use of materials. It is particularly important that the general and interdisciplinary aspects of design be demonstrated and recorded at every stage in the design process. The teaching of design has an integral place in the formation of all engineers, particularly with the increased understanding of the importance of considering sustainability and environmental and social impact.

The Nature of Design Principles

Engineering design can be considered as encompassing three stages. The definition of the need to be met, the conception of a response to that need, and the organisation and management of the delivery of that response. These three areas of activity can be summarised as:

1. The definition of need, requiring the recognition and understanding of the nature of society, of economics, of humanity's needs. The human qualities of reason, compassion, service and curiosity all contribute to the definition of need. **All design begins with a clearly defined need.**

2. The use of creative vision, requiring the ability to think laterally, to anticipate the unexpected, to delight in problem-solving, to enjoy the beauties of mind as well as of the physical world. The ethos within which the problem is being addressed must be understood. **All designs arise from a creative response to a clearly defined need.**

3. The delivery of a solution to the recognised need, requiring the assembly and management of resources and of team members with the necessary skills and knowledge of natural laws, and of the materials and energies needed to effect an efficient and appropriate creative design. **All designs result in a system, product or project which meets the need.**

Design is recognised as an iterative creative process bringing about the development – physical and cultural – of ways of meeting identified needs. The formulation of clear definitions of perceived needs and processes is emphasised and illustrated in the twelve case studies in Chapter 4.

The principles of need, vision and delivery describe fully the role of the designer in all disciplines, not only in the application of the understanding of natural laws and of the analytical processes, but also in the expressed understanding of the societal context. Design is an iterative but controlled process – effective creativity. The need for clarity of definition and control of aims is clearly demonstrated in the examples analysed.

An understanding and knowledge of principle are essential if design decisions are to produce desirable results. In engineering terms a desirable result is likely to be a useful physical entity or system, valuable in meeting some need and improving the quality of life of individuals or of communities.

The Oxford English Dictionary defines a principle as:

- Origin, source; source of action
- That from which something takes its rise, originates, or is derived
- A fundamental truth or proposition, on which many others depend; a fundamental assumption forming a basis of a chain of reasoning.

In engineering it is possible to identify basic principles which can be referred to by all designers of any discipline when initiating their work, or testing the quality of design decisions, whether their own or another's. Such fundamental principles are not to be confused with the postulates, definitions, hypotheses, standards or rules included as part of the technical training and ability of the professional engineer. Such technical formulations are of value in particular instances, and are well known and formulated in engineering practice.

The "principles" presented are not the purely scientific hypothetical principles, such as the laws of statics, dynamics, thermodynamics or electro-magnetism, which are already an essential part of the engineering curriculum. These fundamental principles are intended to provide the total context for good design. They are not necessarily rooted in physics or mathematics, and derive more from experience and practice than from formal theory. They are the substance of professional engineering judgement. An understanding of these basic principles enables an engineer to engage in the highest level of decision-making, to which he can then bring his professional skill and training.

Engineering design involves many parameters upon which the success of the project depends, and each of these specific areas of concern has its own sub-set of laws, standards, practices, codes and regulations. The volume and detail of these particular constraints and directions can make it difficult to appreciate the fundamental principles, which may be well known to experienced designers, but may not have been clearly formulated. The case studies show the relationships and differences between the specific and the general principles by demonstrating these at each step in the design development.

Practice of Design

In practising design the process develops from stage to stage. The factors to be considered at each stage – need, vision and delivery – have been formulated as checklists for use in the early stages of design as follows:

Formulation of Need

- The cultural societal and physical ethos in which the project is to be carried out
- The history of the project, how did the perception of need arise?
- The decision-making structure – social, client, design management
- The human resources available – professional skills, research, available craft skills
- The physical, technical and economic resources available
- Sustainability requirements, energy sources, environmental impact
- Relevant research and development
- **The clear formulation of the need**

Human needs can arise from many sources. They can be recognised politically, economically or socially. The first thoughts may reflect an individual awareness of need, or a professional awareness, linked with a justifiable wish for a fee or contract.

It is of value to consider the several points listed above, and to set these down as they relate to a particular awareness of need. Collectively, they will provide the basis of a "brief" for the project or system, and a review of the context. This will involve an exchange of views with others, but at this stage only a limited number of interested parties will be concerned. The final definition of need will be best expressed in a single, clear sentence.

The designer's skills must be related to social and contextual factors to be effective. He or she is not engaged in abstract research or exploration, but in directly improving the quality of life in society. The care and well-being of the community are of paramount importance, requiring a sympathy and comradeship with others. A sense of humour and the enjoyment of the company of others are of great importance The needs to be met are human needs.

The brief schedule of factors to be considered during this stage of design reflect these human interests and require an understanding of the interaction between others. Our technological skills enable us to serve these interests.

An example of this social dimension was evident in the siting of the University of Surrey in Guildford. The first Vice-Chancellor of the University – Dr. Peter Leggett – was concerned that the university should become part of the fabric of the community, not an isolated institution. We identified alternative sites for the university, one on the slopes to the north of the Cathedral on Stag Hill, and the other on the fields to the west, beyond the motorway. The latter was technically a much better building site, but the former was within a short walk of the city centre. This site was chosen and the technical skills used effectively to stabilise the slope and overcome the building problems.

I recall meeting with Dr. Leggett and discussing the purpose and aims of the university, which we formulated as: "to enable the individual development of the students and equip them to serve the community". The proximity of the city was an important part of meeting this need. The meeting had emotional, caring aspects as well as the formulation of the professional brief. It provided the quality of energy and commitment needed to support the project during the 4-year design and construction programme. Dr. Leggett and I pinned this aim to our study walls for the period of the project development!

Creative Response – the Vision

This stage can be considered as in two parts: the definition and understanding of the social and physical context within which the need must be met, and the development of creative vision of how the need might be met within that context. The key factors of these two sub-divisions of the visionary process are set out below.

Contextual Constraints

- Establish relationships with client team and socially affected groups
- Selection of team with necessary professional competence, knowledge of natural laws
- Surveys of context – physical, climatic, topographical, materials, energy
- Legal requirements – planning permissions, legislation.
- Funding and value for investment, competition
- Programme requirements
- Health and safety issues
- **Summary of contextual items**

The context within which the need is to be met must be appraised. A thorough review of all the factors listed above is essential. This will include physical and social parameters: topographical, economic, climatological, social, historical, *etc.* These should be set down in a contextual report, and referred to during the creative design stage. They may involve the services of specialists (surveyors, geologists, economists, lawyers, planners) working as part of the overall design team.

Care should be taken *not* to attempt to formulate a design before the constraints have been considered. Premature conceptions are frequently abortive!

Effective Creativity

- Interdisciplinary discussions
- "Brain storming" sessions
- Preparation of alternative imaginative schemes
- Programme for delivery
- **Selection of a preferred solution and its clear definition**

This stage of "visionary" design can be the most significant and important of the whole process. The placing of the formulation of the need within the context is the first step. The selection of a suitably experienced, talented and well-trained group of designers, capable of effective interdisciplinary discussions and of working together, is essential. It may be valuable to call on particularly experienced and talented designers who are not part of the main project team, but who can make particularly relevant contributions to the creative design process. The team leader must be a good listener, and sufficiently respected by the group for them to accept his directions.

In 1992 an invitation was received from the Soviet Academy of Science for two senior engineers from the UK Royal Academy of Engineering to visit Russia and advise young Russian engineers on how to work with a market economy. This proved very interesting in increasing our awareness of the contrasts between the centralised communist government and the free market economy. An illustration of this came in a major meeting in St. Petersburg with the senior management of a very large and well-equipped naval research institute, employing about 3000 highly qualified staff. When we suggested that the organisation might be subdivided and liaise more closely with European institutions, the chief executive was very concerned – he asked, "How will I remain in control?! " The delegation of responsibility was not part of their established culture.

It is sometimes helpful to have group "brain-storming" sessions away from the usual office demands. Given good support (flip charts, computers, admin and other support staff) the freedom to work together can be very creative. A number of conceptual designs may be developed and compared. The inspired designer should enjoy the whole process. As an example of the importance and value of being aware of the situations in which we find ourselves, the following note on one of the "inspirations" arising from observation may be of interest.

In designing the terminal works for the Channel Tunnel it was necessary to consider the random movement of high traffic flows with a variety of apparently independent drivers. Observing one day from a high level footbridge at Cannon Street Station the movement of pedestrians leaving a train and descending to an underground pedestrian tunnel, it was clear that whilst they all felt that they were making decisions, they were actually following a turbulent flow pattern from their trains to the tunnel. We therefore adopted a turbulent flow analysis for the traffic arriving at the tunnel terminal and found that it enabled us to develop a traffic management system which works well.

The selection of a preferred solution from the several that may be proposed is very important, and, as was the case with the formulation of the need, it should be clearly defined and well illustrated with good conceptual drawings and diagrams. This vision will form the starting point for the delivery process. It should include a viable budget and programme. It must be acceptable to all the relevant authorities and fully understood by the client body.

The accepted vision can be displayed on the office walls of all those involved in the production and delivery process.

Delivery

- Agreement and documentation of preferred solution – continuous reviews of delivery against aims – the design audit
- Definition of management team and process of delivery
- Selection of key designers, of contractors, sub-contractors, and specialists
- Definition of their roles
- Reference to interested third parties
- Agreement on financing – budget, cash flow, cost control procedures
- Agreement on delivery programme
- Agree contract procedures
- Assembly of project/product resources – skilled crafts, materials, equipment
- Agree quality management procedures – design, manufacture, construct
- Health and safety considerations
- Legal requirements and programme
- **Hand over to clients**

This stage in the realisation of the solution to the need usually takes up the major part of the whole project programme. The delivery process will also be the period of greatest expenditure.

The selection of an experienced and strong project team is essential, capable of bringing together the many specialists involved in the process and of ensuring that the accepted design is not allowed to be changed for short-term expediencies.

Collaborative work is essential in an emergency. During the construction of the oil refinery at Grangemouth in Scotland there was a major incident one weekend when there were no workmen on site. The chief contractors engineer and the young supervising engineer spent several hours excavating in some foundations threatened by settlement. The safety of the work was more important than their respective professional roles.

It is important to set out clearly the project structure and the relationships between the various team members, particularly those between the different disciplines (client, user, lawyer, politician, planner, economist, architect, engineer, *etc.*) and to emphasise the formulated need being considered and the resources necessary and available to meet this need.

Early site experience included night shift working on a major harbour project. The team worked together very closely on the driving of very large piles. When the piling rig broke down in the early hours of the morning the whole team went to their favourite pub and persuaded the publican to open at 2.30 am. As a very young supervising engineer I was included in this invitation, and was impressed by their collaborative concern for each other's safety, and for the quality of the work they did.

In developing a design it is essential to keep in mind these aspects of the process and to work in a systematic manner, with regular reviews of each stage to ensure that all aspects are considered. Under the pressure of programme and cost control, we may occasionally move forward and not cover all these points, leaving us with a less satisfactory conclusion.

It is usually advantageous to begin the delivery with presentations to all involved – the contractor's work force as well as senior management. It is encouraging to experience how interested all participants are in learning the history and creative processes that have produced the project. In some projects it can also be very helpful to inform interested third parties of the proposals: neighbours, local services, *etc.* This is indicated in the case studies in Chapter 4 of this book.

When beginning the construction of the University of Surrey in Guildford the contractors and the consultants gathered together the total work force, and we explained to them the basis of the design, and what we hoped to achieve. We then found that we were spoken to frequently during our visits to the site by various workmen, with suggestions as to how the design might be improved in detail. This was very encouraging and gave us the feeling of operating as a complete team in the construction.

The ethical aspects of design: are of great importance. The changes to all those affected by the work and by the environmental impact need to be carefully considered and noted in the documentation of the developing designs.

Satisfaction of a difficult job well done. The construction of the major airfield in the Falkland Islands was a very demanding project. On the completion of the main runway on time the whole construction team lined the airstrip to greet the first flight in. The young site engineer was very moved to see these very tough workers celebrating this major achievement in tears.

Performance in Practice

With significant projects it is valuable to review their performance in practice.

This can suggest possible post-contractual improvements, and serve to make the users of the project aware of the considerations that led to its completion in a particular form. Such reviews can also be of value in other projects and add to the database of experience available to designers. The factors to be considered during a performance review are:

- Reassessment of design brief
- Possible design development
- Social impact assessment
- Client/user satisfaction
- Economic performance
- Maintenance/operational factors
- **Experience to incorporate in future projects**

The case studies given in Chapter 4 are set out against these principles and show how they apply both to very large and technically complex projects and to the simplest of projects, from the Channel Tunnel to individual but effective projects such as artificial limbs. The criteria are included in an appendix for use in practice.

Example of Case Study

A study of an apparently very simple example of decision-making is given below to illustrate the design process. Using the criteria for each stage as developed above, the process is shown for the selection of new chairs for a college in London.

Formulation of Need

- The cultural societal and physical ethos in which the project is to be carried out
- The history of the project, how did the perception of need arise?
- The decision-making structure – social, client, design management
- The human resources available – professional skills, research, available craft skills
- The physical, technical and economic resources available
- Sustainability requirements, energy sources, environmental impact
- Relevant research and development
- **The clear formulation of the need**

A private college in London is considering the acquisition of new chairs. It is important that the need is clearly understood before committing to the purchase of the many chairs needed. The college has some distinctly differing needs:

- There are about 250 students attending the school every evening and at weekends.
- There is a refectory which may be dealing with some 150 students at a time, but the use of chairs here is for short periods only.
- There is an attractive entrance hall with an adjacent library, and about 20 chairs are needed for these public areas.
- There is a large meeting hall which seats about 100 people, used occasionally.
- There are 20 classrooms seating about 20–25 students each.

Considerable mobility of chairs is required to suit the variable student needs.

It must be possible to stack chairs efficiently when not in use. Some variations in height may be desirable. All maintenance is by voluntary workers, and must be simple and time efficient. Consideration must be given as to whether some chairs with arms are needed. Since students use these chairs for about two hours every evening, they must be comfortable. The students' ages vary from teenagers to octogenarians. About 300 chairs may be needed. Economy is of great importance.

Formulation of need: 300 chairs are required, easy to move, comfortable, attractive and flexible in their usage. They must be easy to maintain and durable.

Creative Response – the Vision

Contextual Constraints

- Establish relationships with client team and socially affected groups
- Selection of team with necessary professional competence, knowledge of natural laws
- Surveys of context – physical, climatic, topographical, materials, energy
- Legal requirements – planning permissions, legislation.
- Funding and value for investment, competition
- Programme requirements
- Health and safety issues
- **Summary of contextual items**

The specification must be agreed with the college organisers and user representatives. The number of chairs and their distribution must be agreed, and possible variations in type to suit the differing needs. Enquiries will be made to obtain information on the range of chairs available on the market. It may be helpful to consult experienced interior designers, and possibly to seek the experience of other similar colleges. Chairs available will be considered relative to the design and decoration of the various rooms: refectory, hall, large meeting room, and study rooms. Suppliers will be invited to tender, and to quote delivery programmes.

Effective Creativity

- Interdisciplinary discussions
- "Brain storming" sessions

- Preparation of alternative imaginative schemes
- Programme for delivery
- **Selection of a preferred solution and its clear definition**

The collected information and specifications will be gathered together and presented to a selection group of experienced tutors and administrators. Based on this discussion a decision will be made as to the chairs to be purchased. This will include cash flow and delivery programmes agreed with the College Treasurer. This will form the instructions for the delivery team, to be confirmed as acceptable by the suppliers.

Delivery

- Agreement and documentation of preferred solution – continuous reviews of delivery against aims – the design audit
- Definition of management team and process of delivery
- Selection of key designers, of contractors, sub-contractors, and specialists
- Definition of their roles
- Reference to interested third parties
- Agreement on financing – budget, cash flow, cost control procedures
- Agreement on delivery programme
- Agree contract procedures
- Assembly of project/product resources – skilled crafts, materials, equipment
- Agree quality management procedures – design, manufacture, construct
- Health and safety considerations
- Legal requirements and programme
- **Hand over to clients**

Several of the items in the checklist above will have been considered in the design phase, but will need confirmation by the suppliers. The detailed programme for delivery will be agreed to ensure minimal interference with the working programme of the college. This may be best done during the vacation period, and may include the disposal of the existing old chairs. A careful check will be kept on the quality of the chairs delivered.

The cash flow programme will have been agreed with the chosen suppliers and the College Treasurer.

Performance in Practice

- Reassessment of design brief
- Possible design development
- Social impact assessment
- Client/user satisfaction
- Economic performance
- Maintenance/operational factors
- **Experience to incorporate in future projects**

The need to consider the selection of new chairs has arisen out of a review of the existing stock. After completing the purchase of new chairs it is necessary to review the brief, and assess the specification for possible future chair replacement. The use of the chairs will be noted, and the opinions of the users obtained.

Summary of Design Procedure

- Agree programme with client, and the scope of work of the designer
- Assess and define need, checking against the basic criteria
- Review and define the context, arrange any necessary surveys, *etc.*
- Carry out as a team the creative design process
- Agree and define accepted design solution
- Clearly define delivery team and programme, including contractors/suppliers
- Monitor progress on a regular basis, with formal review meetings

3 Case Studies – Introduction

 Statements of Principle

The essential initiating activities of all design projects have been set out and discussed in Chapter 2. The three stages of need, vision and delivery, with some additional notes on the performance of the project, are used in the analyses of 12 case studies from a variety of engineering projects.

The design concepts of each project have been described in showing how the projects were conceived, how the design solutions were reached, and how the delivery procedures contributed to the realisation of the project. A schedule of the projects considered is given below.

Material needs

1   The development of a high-throughput automated analyser, winner of the RAEng McRoberts Award 2003
2   The design of an artificial joint

Social needs

3   The Channel Tunnel Terminal, Folkestone
4   Design of the Channel Tunnel Rail Link, Folkestone, London
5   The design of the London Eye
6   Lesotho hydro-electric project
7   Timber Arch Bridge, Melbourne, Australia

Intellectual/Spiritual Needs

8   The design of the Bahá'í Temple, Delhi

9   The design of the Hampden Gurney primary school in London

10 The design of the Bridge Academy secondary school in London

11 The design of residential block for the Harris Manchester College, Oxford

12 The design of the Angel of the North sculpture, Northumberland

 Case Study Format

The case studies are all presented in the same format, using the basic principles of need, vision, delivery and performance. This format is also followed in the proposals for a design audit in Chapter 5. The activities appropriate to each of these fundamental areas of activity are as set out in detail in Chapter 2. A brief statement of the three basic principles is repeated below.

Need – All Design Begins with a Clearly Defined Need

This first principle requires the recognition and understanding of the nature of society, of economics and of humanity's needs as set out in Chapter 2. Reason, compassion, service and curiosity all contribute to this very important first step in design.

Vision – All Designs Arise from a Creative Response to a Need

This second principle is the conception and management of a creative vision to meet the need. It requires the ability to think laterally, to anticipate the unexpected, and to appreciate the aesthetic and emotional aspects of problem-solving as well as the material aspect. This stage in design is considered in two parts: the recognition of contextual constraints and the process of effective creativity.

Contextual Constraints

The ethos within which the problem is being addressed must be understood fully – physically, politically, economically and socially – as well as technically.

Design development is an iterative process. Good relationship within the need-defining team are essential. The perceived needs may change during this stage. Evaluation of the concept requires a full understanding of the need as formulated and the delivery constraints likely to affect the design formulation. The designer also needs to know about market constraints and the production processes.

Effective Creativity

The controlling team or individual must have access to all necessary specialist advice. On larger-scale projects, the management of the various inputs must be strong and effective, but must not inhibit creative thinking. On smaller projects, good self-discipline is necessary to ensure that the development does not deviate from the initial perception of need. External advice must be well coordinated, and its role in the design development understood. Specialist consultants must appreciate the total context and aims of the project, which should not be confused by individual disciplinary objectives.

Delivery – All Designs Result in a System or Product That Meets the Need

The final principle involves delivering a solution to the recognised need in accordance with the vision of how that need might be met. This requires the assembly and management of all the resources and team members needed to complete the task.

As the scale and complexity of projects increase, so does the need to define a clear management structure and to understand the relationship of the design components to the whole project. There should be regular team audits to ensure continuity of concept, as well as applying the controls necessary to ensure a high and consistent quality in the end product.

Performance

A general review of performance highlights aspects of the project and process from which future designs can benefit. It is of value to convene a review meeting of the senior project designers and the operators of the finished project to consider the lessons to be learnt for future similar projects. This review might be best done some 5–10 years after completion.

Summary of Design Procedure

The notes below set out the basic programme for the processes of design adopted for each of the case studies in Chapter 4.

- Agree programme with client, and the scope of work of the designer
- Assess and define need, checking against the basic criteria
- Review and define the context, arrange any necessary surveys, *etc.*
- Carry out the creative design process as a team
- Agree and define accepted design solution
- Clearly define delivery team and programme, including contractors/suppliers
- Monitor progress on a regular basis, with formal review meetings
- Design review after completion and a period of operation

 # Key Features of Specific Case Studies

The case studies in Chapter 4 have been chosen to give a variety of aspects of the engineer's work. These are:

- Automated Blood Analyser
- Intelligent Prosthesis
- Channel Tunnel UK Terminal
- Channel Tunnel Rail Link (CTRL)
- London Eye

- Lesotho Hydro-electric Project
- Maribyrnong Footbridge
- Bahá'í Temple – Delhi
- Hampden Gurney Junior School
- Bridge Academy
- Harris Manchester College, Oxford
- Angel of the North

4 Case Study 1:
Automated Blood Analyser

The subject of this case study won the prestigious Royal Academy of Engineering's MacRobert Award in 2003 for innovation in engineering. It was developed by Randox Laboratories of Northern Ireland. The key elements of design are illustrated through a series of project stages; these culminated in a product being manufactured and sold around the world.

The automated blood analyser has many unique design aspects:

- Considerable scientific challenges had to be met and overcome to prove its viability as a product concept.
- The health-care market is highly regulated in terms of standards and requirements.
- Some of the potential competitors are very large organisations.

The heart of the evidence analysis system is the biochip. This little device can be produced in millions, is used once and then disposed of.

Supporting the biochip is a specially developed instrument to process and read it. This is manufactured in relatively small numbers, but has a working life of many years.

Although ultimately a manufactured product, this type of design can be seen to follow the same need-vision-delivery sequence. There is a collective need which is expressed by the marketplace; eventually, there is delivery of a product which continues over an extended period of time to satisfy the need.

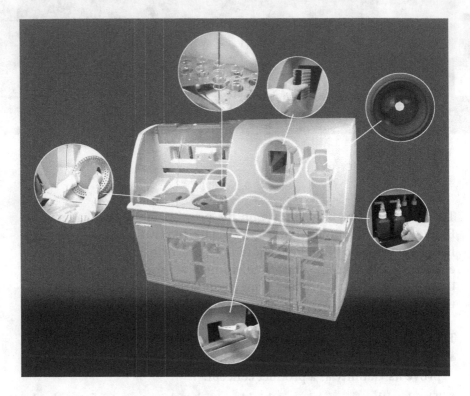

Figure 1. Completed analyser

 # Formulation of Need

- The cultural societal and physical ethos in which the project is to be carried out
- The history of the project, how did the perception of need arise?
- The decision-making structure – social, client, design management
- The human resources available – professional skills, research, available craft skills
- The physical, technical and economic resources available
- Sustainability requirements, energy sources, environmental impact
- Relevant research and development
- **The clear formulation of the need**

Analysis of blood and other bodily fluids is a common procedure in clinical diagnostics. Samples are usually taken from a patient in a doctor's surgery or in a hospital ward, and then sent to a laboratory.

Since 1960 it has been possible to determine the concentration levels of many important substances in blood; this is used to detect the presence of disease, malfunctioning of bodily organs and general level of health. This is usually based on immunoassay techniques.

An immunoassay uses the natural affinity for some molecules (antibodies) to attach themselves to specific proteins (antigens). The degree of selectivity is extremely high; antibodies can be developed or created to attach themselves to a particular protein, and not to others. By linking either the antibody or antigen to a physical phenomenon such as chemiluminescence, it becomes possible to measure its concentration, and hence infer the concentration of the other molecule.

Each substance requires a unique test; the doctor will request whatever tests are needed. The information which comes back can be of vital importance in the diagnosis and in determining the best treatment.

Improved information can be beneficial for the patient and organisational efficiency; however, the consequences of misdiagnosis can be very serious. Although millions of tests are performed each year, clinical diagnostics represent only a small fraction of the total world healthcare expenditure (which is over $300 billion).

There can be some delay (hours or days) until the test results are returned. The diagnosis may need some refinement, with a request for further tests; this would require fresh samples from the patient. This leads to additional work, patient visits and time.

Randox were in the business of supplying diagnostic kits and reagents for these types of tests and had considerable experience in the field. Whilst considering how the market could develop for the future, the possibility of performing a whole set of tests on a single sample seemed worth investigating. If the sample could be kept small and the tests kept independent of each other, then this approach could revolutionise clinical testing.

There have been several technological advances in the industry over the years, but recent attempts to introduce radically new forms of diagnostic testing have been only moderately successful; the technical difficulties of establishing a stable and repeatable system are significant. When moving towards nanotechnology, an increased degree of control is required over the biochemical reactions involved.

The company's upper management included significant scientific experience and understanding. Not only did this greatly assist the task of assessing strategic alternatives and technical risks, but also contributed directly to the innovative process itself.

Financial and material needs of the project increased as the concept began to take shape. Investment in new resources and new expertise was largely funded directly from the company's turnover. Where this can be managed, there is freedom from the overheads associated with grants and external financial support.

Environmental issues pervade many aspects of such a design project. The challenge of mastering a new and difficult technology must never blind one to the real needs of living with one's neighbours. This could be in terms of hazardous substances being consumed by the product, or used in manufacturing it; other considerations are electromagnetic compatibility of the instrument. Some aspects of the invention are inherently beneficial to the environment, such as the significant reduction in sample and reagent volumes per test.

All of the above considerations are fine, but there still remained an incredible task ahead for those involved in the design. The general concept is of a miniature automated analysis system capable of delivering reliable results from an array of tests. How can this general concept be turned into a world-beating innovation?

The ideas centred on the "biochip". A single sample washed over the chip would react with each test site independently and so produce the battery of test results.

How can minute volumes from a patient sample be arranged for multiple, yet simultaneous tests? Can the individual tests be isolated from each other whilst keeping the dimensions to a minimum? Is it possible to achieve accurate results with such a scheme? Can it be manufactured at a reasonable cost?

This fundamental technology development demanded several lines of critical research and development. Several large players had found it difficult to overcome some of them.

To illustrate the difficulties, consider just one crucial problem that emerged during this early work. The biochip was to have several minute test sites; each required a certain type of antibody to be fixed in position; different test sites would have different antibodies. The technique used to attach them to the bio-chip surface must not interfere with their ability to capture and bind other samples.

This required extensive research into surface chemistry as well as binding techniques (such as the use of separate molecules as linkers). All of this is on a microscopic scale, with the demands of stability and lifetime, as well as economy.

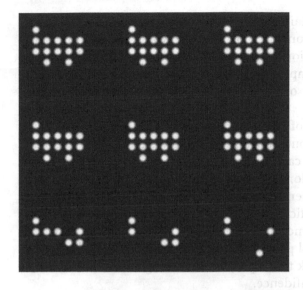

Figure 2. Nine biochip

A clear definition of the need has emerged: To provide a world-class cost-effective system for simultaneously delivering multiple tests from a single patient sample.

 # The Creative Response – the Vision

Contextual Constraints

- Establish relationships with client team and socially affected groups
- Selection of team with necessary professional competence, knowledge of natural laws
- Surveys of context – physical, climatic, topographical, materials, energy
- Legal requirements – planning permissions, legislation.
- Funding and value for investment, competition
- Programme requirements
- Health and safety issues
- **Summary of contextual items**

The client team for manufactured products such as this can be found in the form of the marketplace, or the hospitals and clinical laboratories around the world. The main socially affected group are patients; however, a moment's reflection will show that the whole community is affected by such an innovation. Working practices may alter, and the general level of health may improve. How and where manufacturing occurs could affect individuals or communities, either beneficially or adversely.

Guiding the evolution of some fundamental aspects of the invention was a clear sense of the context within which it was to be used. In broad terms, this is the health care system of the world. Existing practises, cost pressures, organisational structures and workflows for this type of analysis are all relevant here. Any unusual demands by the new system on infrastructure or practices would require critical consideration.

As the concepts became more firm, costs rose. Carefully directed funds allowed the general project to move forward, but also dealt with any areas of potential risk to the project. Analysing and reducing these risks greatly improve confidence.

The clinical diagnostics market is very competitive; any of the large world-players would love to develop new technology and systems that change the market. When it was recognised that a good concept had been born, the choice for the company was to take a risk and move the technology forward, or else sit back and watch the competition do the same.

What was visualised was one year of initial technology feasibility studies and development; this was to be followed by a year developing working prototype systems. A third year was needed to see manufactured systems on the market with a few key tests available on biochips. As systems were delivered, further development was to introduce additional tests to broaden its utility.

Also foreseen in these early days was the need for an extensive and exacting programme to bring this technology under control. The world's regulatory organisations and the potential customers all have to be convinced that this product works accurately and reliably and is of clinical benefit. In every area of the biochip's design and manufacture, in the instrument's processing and analysis, significant innovation was required to establish it was a quality product.

Effective Creativity

- Interdisciplinary discussions
- "Brain storming" sessions
- Preparation of alternative imaginative schemes

- Programme for delivery
- **Selection of a preferred solution and its clear definition**

Although there was an exciting idea for an innovative product, a little reflection indicated there were significant areas of uncertainty. Any one of several areas could have made the project unviable. In these situations, it is vital to plan on reducing areas of risk as soon as possible. A couple of examples will be used to illustrate this. The first deals with the question of technical feasibility of detection; can reactions on such a small scale be measured? The second deals with biochip manufacturing practicality; is it possible to realistically manufacture such a device?

Detection Feasibility

Reduction in the volume within which the reactions take place has the unfortunate effect of reducing the magnitude of any results; accurate and reliable measurement becomes more difficult. Is there a risk that such techniques are just not possible at a realistic cost?

A review of techniques for converting biochemical reactions to physically measurable phenomena included chemiluminescence. The results of the reaction are made to glow; the stronger the result, the brighter the light output. Although there are other techniques such as radioactivity, there are many advantages of using this, environmental concerns being one of them.

With several individual test sites (DTRs) in each biochip and several biochips in a carrier, it was worth investigating the possibility of utilising a camera to look at all simultaneously, rather than using a single detector to measure each in turn. This would allow all reactions to be initiated at the same time and detected at a certain time later.

The challenge here is to ensure the measured light is predominantly due to the reaction and not to side effects or stray light. With most cameras, a limited ability to see very dim objects is also an issue. A special low-light camera had to be utilised, with cooling to tens of degrees Celsius below zero being required. Generally, surfaces were made black and non-reflective to ensure only light from the reaction was detected.

One near oversight was the meniscus on top of the liquid in which the reaction was taking place. Due to the small dimensions involved, most of the liquid surface is curved which leads to a lens effect and distortion

Figure 3. Hand set and carrier

of the image seen by the camera. This was overcome through careful shaping of the reaction well.

These and many more aspects of detecting extremely small quantities of reaction product had to be researched; a mixture of paper review, theoretical analysis and experiment was necessary to realise the optimum approach, and then demonstrate it was suitable for the intended product.

Biochip Feasibility

The general idea of performing multiple tests on a single biochip puts several demands on the reliability and accuracy of its manufacture. Is there a cost-effective way of achieving this?

Aluminium oxide was selected as the best material for the biochip. Surface activation or preparation of the biochip to receive the ligand molecules was performed in various ways.

A programme of work was commenced to determine the optimum means of attaching the ligand (or binding) molecules to the biochip surface, as well as the optimum surface material. Attachment had to avoid affecting the properties of the ligand molecules, on which the whole process depends. Isolation between the various discrete test regions had to be guaranteed to avoid confusion of results. Uniformity across the whole chip was also of great importance.

Analytical techniques established the molecular and atomic characteristics of the surface at each processing stage. This gave confidence that the surface was being modified in the expected way. Gradually, the optimum methodology was established.

It was now necessary to establish the roadmap from a technically sound prototype to a commercial reality. The design is to be transformed into a controlled manufacturing environment. Infrastructure has to be developed to ensure the product can reach the marketplace and be fully supported. On the way there are several regulatory issues that need to be faced.

Representation from manufacturing, marketing, quality assurance are all actively involved at this stage, their respective views combining to create a coherent programme for delivery.

For the biochip, special expertise was necessary to consider how millions of biochips could be made whilst maintaining stringent quality control. To be launched with a small number of biochip-based assays, the same considerations would be carried forward to new assays developed later on. Some of the technology would have to be unique and specially developed; elsewhere it was necessary to carefully select the most cost-effective manufacturing equipment.

For the instrument, it was necessary to embed strict quality controls into its liquid and software processing. At every stage, checks were to be incorporated to ensure that limits were not exceeded, non-linearities were managed, and unexpected results were detected.

Careful consideration was given to the human interface; here is how any user perceives the system, how it is used, and what it "feels" like. If this is managed well, users will like it; initial responses have a lasting effect.

Technology changes so rapidly that it is necessary to look forward in the direction it is heading. Selection and proving a particular technology can take some time, and it could be "old" before there has been much product out of the door.

▦ Delivery

- Agreement and documentation of preferred solution – continuous reviews of delivery against aims – the design audit
- Definition of management team and process of delivery
- Selection of key designers, of contractors, sub-contractors, and specialists
- Definition of their roles
- Reference to interested third parties
- Agreement on financing – budget, cash flow, cost control procedures
- Agreement on delivery programme
- Agree contract procedures
- Assembly of project/product resources – skilled crafts, materials, equipment
- Agree quality management procedures – design, manufacture, construct
- Health and safety considerations
- Legal requirements and programme
- **Hand over to clients**

As the design matured, each of its facets saw its own optimum solution settle; the overall coherent design began to emerge. The enormous amount of documentation necessary to orderly transfer the design into a controlled manufacturing environment took place. Advances were also made in clarifying the techniques to be employed for manufacturing.

Early consideration of regulatory approval formed an essential aspect of the project plan. Approvals can take many months, requiring significant documentary evidence of compliance and some external testing for each of US, Europe and Japan.

An intricate set of complementary suppliers and sub-contractors, work instructions, quality acceptance criteria and many more key items had to be established. Every aspect of the product, how it is to be procured, fabricated, assembled, tested, quality-assessed, packaged and distributed, needed to be carefully considered and settled.

Most processes were outsourced, but some were determined as best undertaken in-house. With some critical supplies and some key areas of technology, special efforts were made to establish dual-sourcing; if one source failed, there needed to be a back-up.

Earlier estimates of product volumes determined the most suitable technology to use; more detailed analysis was now employed to minimise initial product costs.

Capital cost considerations kept a brake on some areas of significant expenditure until it became strictly necessary. This required careful project management; to move ahead on any such item, it was essential that all factors necessary for the go/no-go decision were available in good time.

The largest such decision was to initiate manufacturing; from that point on, expenditure would increase significantly. Sales and marketing had determined how many biochips of which type, how many systems, which market areas and when, for the next few years. These refined estimates determined the investment required to launch the product, and the payback period. The product launch date took account of the various trade shows that would best be used to proclaim "evidence" to the world.

After the decision to manufacture, meetings between engineering, manufacturing, quality assurance, marketing continued with a new momentum. The company had to exert significant efforts to get through the labour pains and hopefully healthy delivery of the new baby. Manufacturing now called the shots; critical suppliers and main sub-contractors had contracts finally negotiated and agreed. Patents and non-disclosure agreements (NDAs) had already safeguarded the company's proprietary information.

The company's existing distributors and sales offices around the world were consulted, sharing with them the vision of this major product launch. Test sites were established; these would take early systems and put them through their paces, usually publishing their results. This would ease entry into what can be quite a conservative market.

Discussions with third parties made it apparent that drugs-of-abuse were a rapidly growing market segment that was ideally suited to the "evidence" system. This became one of the first assays to be launched. Plans had also been put in place for the development to introduce additional assays.

After-sales support, both for biochip use and the instrument with its on-site maintenance, had to be arranged. This needed a third-party organisation capable of managing support of such a complex instrument.

As each manufacturing sub-contractor is brought on-line, their contract was carefully dovetailed into the latest company forecast of production volumes. Planning, trial runs, tests, verification all needed to be done for each aspect of the product. Production equipment, clean rooms, all had to be brought into use according to the master schedule.

One of the key requirements is to ensure the developed system is proven to work in such a way that would be recognisable to the medical community. This involves critical analysis of key aspects of the design, as well as final verification and validation. Verification is the demonstration that the system meets the design specification, whereas validation is the demonstration that the system meets the user's requirement.

Widespread acceptance was obtained by getting approval from FDA (US government Food and Drugs Administration). This would have involved inspection of the company's in-house systems and organisational structure, as well as the veracity of any claims made for the new instrument and biochip.

As this was a new venture for the company, insurance cover needed to be carefully reassessed. What if there was an error in diagnosis, based on tests on a biochip? How is the company to be protected?

 # Performance in Practice

- Reassessment of design brief
- Possible design development
- Social impact assessment
- Client/user satisfaction
- Economic performance
- Maintenance/operational factors
- **Experience to incorporate in future projects**

All through the project, an eye was kept on the market and its evolution. As each of the critical processes came under control, this confirmed an aspect of the project's viability in the context of this moving scene. Early trials brought in invaluable information that could feed through to late product refinements. At the actual product launch and its aftermath, all eyes and ears were on feedback from users and reviewers; this was an exciting time!

New test assays were keenly discussed; hitting the right market would accelerate "evidence" onto the world stage. As each new test was launched, the systems around the world received a remote software update to enable the new test.

Good working relationships were established with some test houses. They wished to evaluate the "evidence" system for use in certain diagnostic areas such as drugs of abuse. This resulted in scientific papers being published which were welcome additional publicity.

Feedback was obtained from in-house manufacturing issues as well as users of systems around the world. Sales figures of biochips and systems were analysed, compared with predications, and revisions to budget made accordingly. Quality reviews assessed any areas of sub-optimal yield in biochip production, leading to several improvements. Production costs were continuously analysed, and moves made to reduce them. User surveys led to minor changes being made to the design and software updates to the system.

These are still relatively early days; the vision of better patient testing with improvements in diagnostic efficacy has been realised in certain areas of clinical diagnostics. New assays continue to be developed as interest in "evidence" grows and its market develops. Feedback from manufacturing and users is still feeding improvements in quality and manufacturing technique.

The excitement of those early days of product concept continues to awaken new possibilities in the clinical diagnostics market, benefiting patients and health services alike.

4 Case Study 2:
Intelligent Prosthesis

 Formulation of Need

- The cultural societal and physical ethos in which the project is to be carried out
- The history of the project, how did the perception of need arise?
- The decision-making structure – social, client, design management
- The human resources available – professional skills, research, available craft skills
- The physical, technical and economic resources available
- Sustainability requirements, energy sources, environmental impact
- Relevant research and development
- **The clear formulation of the need**

The process of design and development of most medical products starts a long while before the final definition of the need. The process of discovery and understanding of the need of the user, in this case an amputee, is rooted in years of research in amputee locomotion. Studying the way the amputee walks and interacts in daily activity, the changing social needs, requirements of different ages, aspirations for rehabilitation and survival in different socio-economic structures are only a few of the topics of general consideration.

This case study arises from the introduction of microprocessor technology in an environment governed by mechanical devices. It covers the development of products which are connected to humans who are impaired by loss of a limb. The project describes the development of a prosthetic foot.

Advances in medical science and improvements in the rehabilitation of lower limb amputees using standard affordable working mechanism are aspects of facilitation and delivery which enable the needs to be met more effectively.

The aims of the manufacturer require that an alliance exists with the users and practitioners to give a sense of ownership by all the "stakeholders". It is important that there is a shared recognition that the need moves with the times, recognising competition and focusing on areas of growth in value.

The need was to develop a prosthetic foot for amputees, enabling normal activities by patients.

The Creative Response – the Vision

Contextual Constraints

- Establish relationships with client team and socially affected groups
- Selection of team with necessary professional competence, knowledge of natural laws
- Surveys of context – physical, climatic, topographical, materials, energy
- Legal requirements – planning permissions, legislation.
- Funding and value for investment, competition
- Programme requirements
- Health and safety issues
- **Summary of contextual items**

The study of amputees shows a variety of needs. Young amputees require the ability to run, participate in sports, be attractive, yet in some parts of the world the need of the amputees are to work, and they have to compete with able-bodied people. This requires that the artificial limb reduces the effort of walking and enables longer life for the product.

Modern foot design showed the need for independent control of heel deflection. Four active trans-tibial amputees and two trans-femoral amputees were tested using the gait laboratories at King Health Care and Queen Mary Hospital NHS trust in UK. Studies carried out on amputees' ability to go around corners and manage obstacles demonstrated the improvement provided.

Surveys of existing modern feet users provided an insight into the overall function. Additional information on ease of alignment, feeling of load transfer, the sensation during standing, as well as cosmetic appearance and overall impression was collected.

The results of the survey highlighted the issues as perceived by the amputees. Through interviews conducted with five prosthetists on the fitting and selection and matching of the prosthesis, additional needs were identified.

An existing organisation with experience of prosthetic design defined the need for this project, and explored the possible ways to meet the need.

In the case of Blatchford's, the designers and manufacturers of the prosthesis, this involved extending their activities from making and repairing conventional artificial limbs to an engineering operation capable of handling advanced casting and machining and complex composite construction, using state of the art technology and procedures. The amputees' needs and the participation of clinicians and physiotherapists was important in the refinement and verification of the assumption, and helped to minimise and manage risk.

The introduction of microprocessor control was based on the need of the amputee to vary the walking speed. The computer control swing phase device was developed to change its characteristic to suit different walking speeds. This is illustrated in this case study of a new Elite foot.

The Multiflex ankle by Blatchford was pioneered using rubber as part of the main structure of the Endolite system. The use of compliant material facilitates adaptability to uneven ground by allowing controlled motion. The combination of this energy-storing foot with the Multiflex ankle provided a high degree of compliance, providing an energy-efficient gait for most amputees.

The 1990s saw the revolution of many advanced prosthetic foot designs. In many new applications of technology, the solution for delivery requires collaboration with skills available outside the original organisation, achieved through the establishment of relationships with other organisations.

The need includes the identification of future safety regulations, as well as quality assurance and compliance with good manufacturing. In order to provide successful products, it is imperative that users' needs be satisfied. There is also a need to demonstrate that the original design specification remains valid and is met by the prototypes in practice.

Effective Creativity

- Interdisciplinary discussions
- "Brain storming" sessions
- Preparation of alternative imaginative schemes
- Programme for delivery
- **Selection of a preferred solution and its clear definition**

The procurement agencies in most countries like to see the clinical evidence supporting the claims that manufacturers make. In order to meet these, the process of validation in the market at every stage of development from feasibility through initial prototype to production is undertaken.

Validation in the field is complemented by the collection of evidence from kinetic measurements validating the aims of the design. Initial gait analysis data revealed a significant difference between the way the trans-tibial and trans-femoral amputee uses the Elite foot. Additional data collected showed the combined vertical and horizontal ground reaction forces were similar to that of able persons.

In addition to the validation and collection of clinical data, it is important to verify some of the engineering assumptions, particularly those that affect the safety and structure of the foot. Design Failure Mode Engineering Analysis (FMEA) developed for medical devices was used throughout the development stages of the Elite foot. The FMEA outcome showed all the high-risk areas were dealt with by developing specific functional tests to ensure reliability and robustness.

Following analysis of the needs and detailed study of the biomechanical requirements, several solutions were created through a series of brain-storming sessions. These were reviewed, and those which had potential to provide realistic solutions were developed. The vision was set around making the best use of affordable and proven technology and design, combined with the optimum use of advanced materials in the understanding of the science of amputee locomotion. The integration of design and advanced manufacturing enabled the stretching of the boundaries of the application of technology within a realistic framework of mass-produced capability.

A concept was evolved which provided a solution to the main requirements. Once the concept was approved, the product design specification was created as identifying all the requirements.

Stages

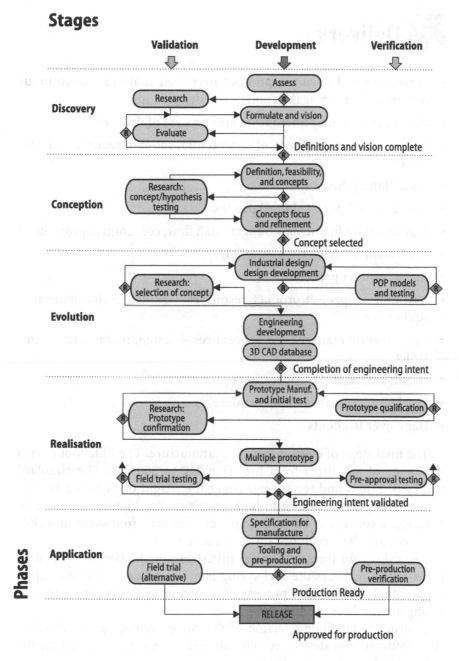

Figure 4. Diagram of need to be met for intelligent prosthesis

■ Delivery

- Agreement and documentation of preferred solution – continuous reviews of delivery against aims – the design audit
- Definition of management team and process of delivery
- Selection of key designers, of contractors, sub-contractors, and specialists
- Definition of their roles
- Reference to interested third parties
- Agreement on financing – budget, cash flow, cost control procedures
- Agreement on delivery programme
- Agree contract procedures
- Assembly of project/product resources – skilled crafts, materials, equipment
- Agree quality management procedures – design, manufacture, construct
- Health and safety considerations
- Legal requirements and programme
- **Hand over to clients**

The final stage of development is manufacture. The Elite foot covers the range of amputees from foot size 240 to 300 mm. The standard nine sets of heel and toe springs cover the weight of 44 to 125 kg amputees doing low, medium to high activities. For higher weight up to 166 kg, the spring can be made to order. Cosmetic foot shells and sliding socks are also developed for manufacture.

The validation through clinical investigation showed that for most trans-tibial amputees the heel spring is preferred one level lower than the original design. This process of selection was reviewed and new fitting instructions produced.

In order to facilitate selection of the correct spring, a virtual selection software was developed, enabling the prescribers to match the activity and weight more closely.

Steps were taken to prepare the market for the new technology, using the media to capitalise on the emotional factor of innovation enables the reduction of the risk of rejection through fears of unknown technology.

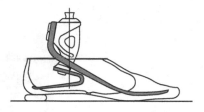

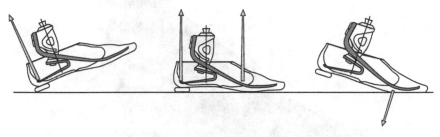

Figure 5. Movement of foot

The shelf-life and product-life were also considered to ensure that they were suitable for the user and in the operating environment.

Other commercial factors such as the manufacturing capability and cost and the possible return on investment were also considered.

 # Performance in Practice

- Reassessment of design brief
- Possible design development
- Social impact assessment
- Client/user satisfaction
- Economic performance
- Maintenance/operational factors
- **Experience to incorporate in future projects**

Despite the longer than anticipated development cycle time, the Elite foot typifies the strength of innovation in meeting today's amputees' needs. The amputees who participated in the early trials have mostly provided a very positive feedback and expressed a perception of enhanced comfort. In one case there is a need to control the heel

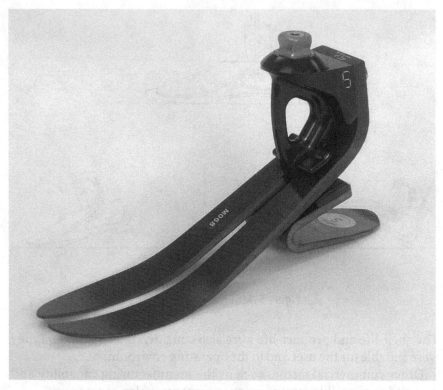

Figure 6. Completed functional joint

function more closely, and in a few early cases, the selection process was a lengthy procedure due to iteration. These feedbacks have been incorporated in changes in the final product, and the only factor not as yet verified is the long-term repeatability and reliability of the product beyond 2 years of life. The early tests complied with a simulated 3–5 years of life. The Elite foot due to its simplicity of design has opened a new chapter for Blatchford product development, where utilisation of the science, and familiarity with the real needs of amputees, enables the production of simple solutions to meet increased demands for mobility. This process has developed new feet for the moderately active amputee, and more advanced feet with a greater capability to absorb greater shock loads, store more strain energy and provide greater propulsion.

Primary biomechanical research will be conducted in the first 6 months of the launch. This includes the use of various devices and measurement of ground reaction forces and moments as well as transfer of loads. Following market assessment and qualitative and technical research, a list of issues to be addressed has been created. Modifications to the design will be generated from this research.

Figure 7. An amputee using the prosthetic foot

4 Case Study 3:
Channel Tunnel UK Terminal

 Formulation of Need

- The cultural societal and physical ethos in which the project is to be carried out
- The history of the project, how did the perception of need arise?
- The decision-making structure – social, client, design management
- The human resources available – professional skills, research, available craft skills
- The physical, technical and economic resources available
- Sustainability requirements, energy sources, environmental impact
- Relevant research and development
- **The clear formulation of the need**

The history of the Channel Tunnel has been recorded in several books. Serious proposals for a tunnel were made first in 1802 by a French engineer. Throughout the next 200 years there were several other proposals; their failure to be realised was partly due to the limitations of the contemporary technical developments, but probably mostly to the political uncertainties of the period.

Following the period of relative harmony and collaboration in Europe after the Second World War and the collapse of the Soviet Union decreasing the "cold war" fears, the increasing confidence in the political stability lead to a wish to improve trading conditions throughout Europe, and the willingness of the UK to form an active component of the European Union. This political and economic optimism, combined with improvements in tunnelling and traffic management technologies, made the possibility of achieving a fixed link between the UK and mainland Europe much more feasible. Forecasts of the traffic growth and the demand to create a viable market contributed to the confidence of the Eurotunnel promoters. An initial move was made in 1970, headed by Rio Tinto Zinc. Scheme designs were developed, but the project was suspended when the government was unwilling to back the project financially, there being competition from the development of the Concorde airliner, and the possibility of a new airport at the mouth of the Thames. The design schemes were archived, until the project was reconsidered in early 1980, and work recommenced under the overall direction of Eurotunnel. A management structure was established with the overall design and building service being provided by Transmanche Link, who subcontracted the major design work.

The assembly of the necessary technical skills and the material resources in both France and the UK, together with the availability of initiating funding from the private sector and increased confidence in the possibility of close international contractual relationships necessary for such a major project, made the project feasible.

A simple and clear formulation of the need became possible: To provide much increased facilities for communication and collaboration by building the best transport facility in the world.

 # The Creative Response – the Vision

Contextual Constraints

- Establish relationships with client team and socially affected groups
- Selection of team with necessary professional competence, knowledge of natural laws
- Surveys of context – physical, climatic, topographical, materials, energy
- Legal requirements – planning permissions, legislation.
- Funding and value for investment, competition
- Programme requirements
- Health and safety issues
- **Summary of contextual items**

The formulation of a design vision of how the project might be achieved in practice required a thorough appreciation of the social, political, economic, and physical context, and the generation of a deliverable design of the complete project within these constraints. Because the project was of national and international significance, it was decided to obtain the necessary authorisation by a Parliamentary Bill, rather than by local planning approval. A scheme design was developed as a basis of the Bill, and objections were heard by Select Committees during the parliamentary procedures in the form of petitions. These procedures took about 6–9 months to complete, evidence being heard in the House of Commons and the House of Lords. The preparation of evidence for the Committees and its presentation by legal counsel for the many interested parties required a great deal of work. The design team was on stand-by for all these procedures, briefing the legal counsel before, during and after the presentation of evidence to the Committees. This required the ability to formulate responses to a variety of questions at short notice.

The petitions varied from simple individual concerns to complex transport and services provisions: from the need to prohibit hang gliding from the escarpment over the terminal site, to the routing and funding of the major road works through Kent.

An example of a minor petition and its resolution.

A violin maker submitted a petition for consideration during the Select Committee procedures. He had moved from being a channel ferry ship chef to becoming a violin maker, because of possible competition from the Channel Tunnel project. He bought a house near the terminal site, which turned out to be between the two carriageways of the motorway entrance to site. He appealed to have the motorway moved, but eventually settled for a new home in a quiet location, where he could tune violins more effectively! He was not cross-examined by the barristers during the Select Committee procedures, for humanitarian reasons!

The terminal site in the UK was an area of "high landscape value" and of "special scientific interest", requiring special approval for alterations to the landscape and environment, in addition to the normal disturbances of any major civil engineering project: compulsory purchase, compensation to existing land owners and users, *etc.* Consideration was given to a considerable variety of issues ranging from major regional transport facilities to the specific personal issues of property owners and the local impact. Compensation was paid to local residents where it was felt that property values had been adversely affected. In addition to the formal legal activities, there were many informal meetings and presentations to local interest groups. Consideration was given to the protection of indigenous floral and insect species. Detailed surveys, topographical, geological, *etc.*, were commissioned to provide a thorough understanding of the context in which the project had to be developed.

The "design and build" contract procedures involved the designers as part of the construction company, with Anglo-French overall management. "Eurotunnel plc" was established as the client body, with Sir William Halcrow acting as independent checking consultants.

The contracting organisation was "Transmanchelink" (TML) with many sub-contractors and consultants, including Building Design Partnership and W.S. Atkins as principle sub-consultants. Extensive economic and legal procedures to set up the companies involved and ensure the availability and control of finance were also necessary. Special attention was also needed to cover the relationships with national transportation networks (road and rail) and the design and provision of specialist rolling stock.

Effective Creativity

- Interdisciplinary discussions
- "Brain storming" sessions
- Preparation of alternative imaginative schemes
- Programme for delivery
- **Selection of a preferred solution and its clear definition**

Having established the need and the context within which the need is to be met, the creative design stage aimed to produce proposals for meeting the need. This required regular meetings of the senior representatives of all disciplines concerned with the design, to ensure a full and shared understanding of the brief and of the need to be met.

The conceptual stage of design involved extended joint sessions of all the key designers with adequate equipment for sketching and presentation (flip charts, *etc.*), free from external interruption. It was found helpful to hold these extended sessions away from the office routine. This design development took place in parallel with the Select Committee procedures, in order to reduce the overall time required. This did

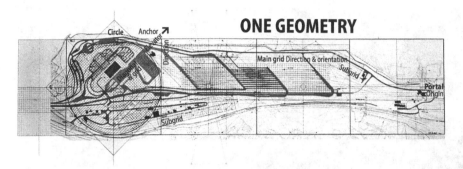

Figure 8. Channel Tunnel Terminal geometry

involve occasional variations in design development to take note of requirements arising in the Select Committee procedures. For example, the main site drainage had been designed to use a balancing reservoir at the edge of the site, but this was disallowed for environmental reasons, and a main outfall to the sea was required – at an extra cost of about £ 2,000,000.

The preparation of alternative imaginative schemes giving an overall view of the possibilities available were summarised as freehand sketches and presentations (Figures 8–11). Frank and uncommitted comparisons of the several overall proposals (free from too much detail) arising from these design sessions were made until full agreement by the whole design team on a preferred solution was made.

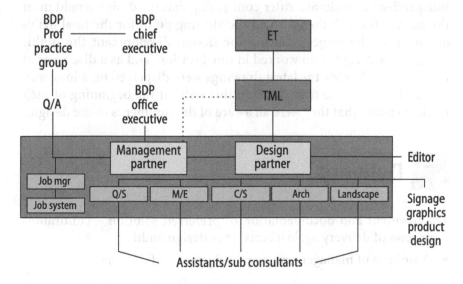

Figure 9. Project organisation

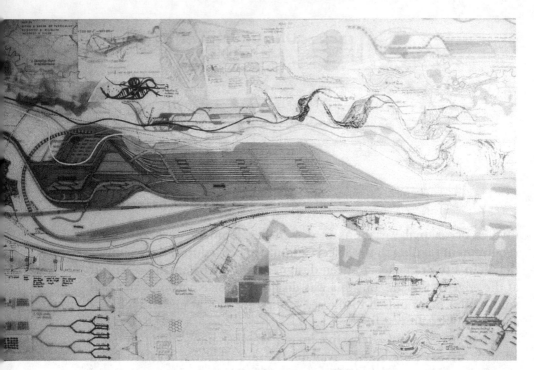

Figure 10. Sketch design

This preferred solution was then clearly defined in words and sketches for future reference and for presentation to the client and the independent consultants. After confirming that the design would meet the need defined, it then formed the starting point for the final delivery stage of the project. During the design development the multi-disciplinary design team worked in one location, and as a discipline at the end of each week the latest drawings were displayed on a long wall in the office, and the team "walked this wall" at the beginning of each week to ensure that they were all aware of developments in the design.

 # Delivery

- Agreement and documentation of preferred solution – continuous reviews of delivery against aims – the design audit
- Definition of management team and process of delivery

Figure 11. Completed project

- Selection of key designers, of contractors, sub-contractors, and specialists
- Definition of their roles
- Reference to interested third parties
- Agreement on financing – budget, cash flow, cost control procedures
- Agreement on delivery programme
- Agree contract procedures
- Assembly of project/product resources – skilled crafts, materials, equipment

- Agree quality management procedures – design, manufacture, construct
- Health and safety considerations
- Legal requirements and programme
- **Hand over to clients**

Management and Control

The delivery stage began with a presentation of the agreed scheme design to the delivery team. The delivery team was defined and the various tasks of the team members established as an overall project management team. The management structure was set out, and the operational procedures, regularity of team meetings, preparation of interim reports, *etc.* were defined.

The overall programme of the various decision and construction stages was established, including post-hand-over maintenance and operational procedures. From this master programme various short- and medium-term programmes and budgets for the various sections of the project were prepared, showing the allocation of responsibilities and identifying critical path events.

These programmes included consultation with other interested parties. Third-party organisations were scheduled and communication arrangements established.

Detailed budgets for the various stages of the project were prepared, as part of a major budget for the whole project. This related to the overall investment programme. Budget review procedures were developed and regular reviews undertaken.

Contract Procedures

The supply chain for consultancy, supply and construction sub-contracts was established. Some contracts were put out to competitive tender, others were negotiated. The capability requirements for all contractors and suppliers were defined, and short lists of possible contractors and suppliers prepared. It was important to confirm the timely availability of labour and materials, and to note the need for any early ordering required to meet delivery targets. Continuous and detailed scheduling of an integrated programme, clearly identifying critical activities, was an essential component of the delivery process. This ensured well correlated activities by all parties. Site quality control procedures covering design, manufacture and construction were established, with the appointment of Clerks of Works and supervisory staff. This included all health and safety procedures, and agreement with external authorities.

Performance in Practice

- Reassessment of design brief
- Possible design development
- Social impact assessment
- Client/user satisfaction
- Economic performance
- Maintenance/operational factors
- Experience to incorporate in future projects

Assessment of Design Brief

The design brief seems to have been satisfactory. The project was completed as set out in the detailed specification and has performed well in practice. The environmental and social impact of the project has been as predicted and seems to be satisfactory. The work approaching completion on the London rail link, whilst rather delayed, will significantly increase the overall effectiveness of the project.

Economic Performance

The cost of the project has considerably exceeded the original estimate. The need for public as well as private funding of such major "public" works needs to be reviewed. The financial charges for borrowed capital have been difficult to meet. Comparison of budget proposals with final project cost need to be made to see what lessons might be learned.

Case Study 4:

Channel Tunnel Rail Link (CTRL)

 Formulation of Need

- The cultural societal and physical ethos in which the project is to be carried out
- The history of the project, how did the perception of need arise?
- The decision-making structure – social, client, design management
- The human resources available – professional skills, research, available craft skills
- The physical, technical and economic resources available
- Sustainability requirements, energy sources, environmental impact
- Relevant research and development
- **The clear formulation of the need**

The Channel Tunnel Rail Link (CTRL) was the first major new railway to be constructed in the UK for over a century and the first high speed railway. The intention was to increase railway capacity between London and the Channel Tunnel and Kent and the rest of the UK. The 109 km of track of the CTRL stretches from St Pancras, central London, to the Channel Tunnel complex at Cheriton in Kent, connecting Britain directly with Europe's expanding high-speed rail network and significantly reducing journey times for passengers, with the potential for freight use if demand requires.

In 1984 an Anglo-French consortium, Eurotunnel, received the concession to finance, build and operate the Channel Tunnel between the UK and France. Financed entirely by the private sector, it opened ten years later.

It was during this period of investment that the Department of Transport's "Kent Impact Study" in 1987 highlighted that whilst the capacity of the existing network between London and the Channel Tunnel would be sufficient to handle international traffic until 2000, thereafter, a new high-speed line would be required to meet the expectations of an increasingly mobile UK and European population, into the twenty-first century.

Between 1988 and 1994 a number of solutions and problems were examined. Finally, a Parliamentary Hybrid Bill was enacted in December 1996. Upgrading the existing track would not have been sufficient to contain all the existing and expanding local traffic, and provide the high-speed link appropriate to the needs of the Channel Tunnel traffic.

The selection of a private sector promoter was launched in 1994, including the amount of government funding needed and the willingness of the private sector to take a risk. The conglomerate company – London and Continental Railways (LCR) – was created as the owners with a lease until 2086. The final project therefore was formed as a PPP, a public private partnership. A team of consultants and subcontractors was assembled to carry the work forwards.

Government grants and agreement to act as a backer of last resort proved necessary to enable the finance to be assembled to carry out the project (total cost £5.2 billion). LCR was created as the parent company to design, build, finance and operate the link; with shareholders who could provide expertise in high-speed railway design, construction management, electrical power supply, transport operation and finance, LCR included as partners Arup, Halcrow, Systra, French Railways (SNCF), Bechtel, London Electricity (now EDF), National Express, Virgin (until 1998) and UBS Warburg. It incorporated into the project the existing BR project team (Union Railways Ltd.) who had spent many years looking at options and developing the project since inception.

Rail Link Engineering (RLE) was assembled as the design and project management team: to form a single organisation grouped together in the same accommodation with personnel seconded from Bechtel, Arup, Halcrow and Systra (the consultancy arm of SNCF).

The Creative Response – the Vision

Contextual Constraints

- Establish relationships with client team and socially affected groups
- Selection of team with necessary professional competence, knowledge of natural laws
- Surveys of context – physical, climatic, topographical, materials, energy
- Legal requirements – planning permissions, legislation.
- Funding and value for investment, competition
- Programme requirements
- Health and safety issues
- Summary of contextual items

The legal and governmental considerations for such a complex and large-scale project are very considerable. These included considerations of the very complex work needed for the London terminal at St Pancras/King's Cross stations, and the connection to the national railway network and the general London transport network. The progress of the Bill through Parliament was subject to the careful consideration in Select Committees of many petitions submitted and attempts made to meet these requirements and compensation given to those claiming to be adversely affected by the project.

The Bill also considered the environmental and regeneration consequences of the construction and operation of such a major project. Listed buildings and other archeological sites were investigated and careful plans made to preserve and enhance these factors. It is estimated that some 50,000 new jobs will be created and some 20,000 new homes built as a consequence of the wider regeneration implications. Much new development of office properties will also be created, particularly in the King's Cross district.

Effective Creativity

- Interdisciplinary discussions
- "Brain storming" sessions
- Preparation of alternative imaginative schemes
- Programme for delivery
- **Selection of a preferred solution and its clear definition**

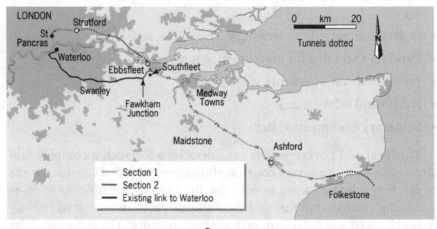

a

Open	Section 1 – opened September 2003	Section 2 – scheduled to open in early 2007
Route	From the Channel Tunnel to Fawkham Junction, north Kent via Ashford International	Southfleet Junction to St Pancras via Ebbsfleet and Stratford
Distance	74 km	39 km
Construction period	October 1998 to Summer 2003	July 2001 to 2007
Construction cost	£1.9 bn	£3.3 bn
Reduction of journey time	20 minutes	A further 15 minutes across the whole route

b

Figure 12. General plan of CTRL route and project stages. **a** Map showing the route of the Channel Tunnel Rail Link. **b** The two phases of constructing the rail link (sections 1 and 2).

Channel Tunnel Rail Link – Overall Programme	
1997 – February	Channel Tunnel Act receives Royal Assent
1988/1990	Joint venture team assembled – Eurorail, Trafalgar House, & BICC
1991	Eastern approach route chosen, giving increased regeneration
1992	Project team set up as Union Railways (URL), includes consultants
1993/1994	Preferred route agreed, naming St Pancras for final approach
1994 – May	Queen officially opens the Channel Tunnel
1884 – November	CTRL Bill goes before Parliament
1996	London & Continental Railways (LCR) awarded/contract
1996 – December	Hybrid Bill receives Royal Assent
1997/1998	Funding of project negotiated, including Governmental support
1998 – October	Construction begins on Section 1
2001	North Downs Tunnel completed, construction begins on Section 2
2003	Test run on Section 1 – speed of 334.7 km/h beating previous UK record of 1979
2003 – September	Prime Minister accepts completion of Section 1, Section 2 50% complete
2006 – June	Track laying and St Pancras station refurbishment to be completed
2007	Section 2 to be opened, CTRL fully operational

Figure 13. Programme

The CTRL was planned to be built in two sections: 74 km from the Tunnel through Kent to the outskirts of London, and another 39 km through the existing railway network on to St Pancras and King's Cross. There are also international stations along its route at Stratford in East London and Ebbsfleet in north Kent, resulting in substantial redevelopment around St Pancras and Stratford (Figure 12). The detailed development of these two stages are given in Figure 13.

Compulsory purchase of the land and property along the route was authorised through the CTRL Act 1996. To reduce the environmental impact, the line intentionally follows existing transport corridors through Kent. The CTRL Environmental Statement that accompanied the CTRL Bill set out how the full environmental impact was assessed.

 **Delivery**

- Agreement and documentation of preferred solution – continuous reviews of delivery against aims – the design audit
- Definition of management team and process of delivery
- Selection of key designers, of contractors, sub-contractors, and specialists

- Definition of their roles
- Reference to interested third parties
- Agreement on financing – budget, cash flow, cost control procedures
- Agreement on delivery programme
- Agree contract procedures
- Assembly of project/product resources – skilled crafts, materials, equipment
- Agree quality management procedures – design, manufacture, construct
- Health and safety considerations
- Legal requirements and programme
- **Hand over to clients**

Figure 14. Detailed layout at King's Cross

Considerable effort was expended to get public support, to seek to inform/reassure/persuade all those living on the proposed route or affected in some other way, and to establish good, dedicated links with all interested parties and set up systems to streamline approvals, *e.g.* with Railtrack, the Highways Agency, local authorities.

As one of the largest and most innovative projects since the Victorian era, Section I was completed both on time and within budget – a significant achievement – being formally opened in September 2003. Section 2 is scheduled for completion in early 2007.

The project consists of :

- 25% of the route is in tunnel
- 60% is within existing road or rail transport corridors

- There are 152 bridges along the route
- Over 50 million man-hours have been worked during the construction

Figure 15. Layout of major works at Ebbsfleet

- 16.5 million m^3 of earth have been excavated
- 11,500 piled foundations have been installed
- 500,000 m^3 of concrete have been poured
- Eight giant tunnel-boring machines were used for the Thames and London tunnels

 Performance in Practice

- Reassessment of design brief
- Possible design development
- Social impact assessment
- Client/user satisfaction
- Economic performance

Figure 16. Medway Bridge

- Maintenance/operational factors
- **Experience to incorporate in future projects**

The St Pancras and domestic stations facilitate easy exchange between inter-city services and Eurostar, with the construction of new platforms for cross-London Thameslink services/main line connection to northern England, Scotland and Wales.

Regeneration is being encouraged along the main international stations on the route. The Thames Gateway area will be boosted by additional developments worth about £500 million. An estimated 50,000 new jobs will be created in East London and the Thames Gateway with some 20,000 new homes and some 20 million sq.ft. of office and business development. The Rail Link Countryside Initiative (RLCI) was set up as an independent charity to enable local communities and organisations to realise their ideas for environmental enhancement along the route.

Whilst ultimate ownership will remain with the government, LCR will have a lease for the track until 2086 and the commercial opportunities created. There are close contractual and organisational links between the LCR and the Eurostar train services.

4 Case Study 5:
London Eye

 Formulation of Need

- The cultural societal and physical ethos in which the project is to be carried out
- The history of the project, how did the perception of need arise?
- The decision-making structure – social, client, design management
- The human resources available – professional skills, research, available craft skills
- The physical, technical and economic resources available
- Sustainability requirements, energy sources, environmental impact
- Relevant research and development
- **The clear formulation of the need**

Throughout the 1990s there was general public and official enthusiasm in the UK for the construction of significant and novel projects as a means of celebrating the coming millennium. This enthusiasm created the opportunity to attract the finance, either public or private, necessary to realise attractive projects which were otherwise unlikely to be built.

The London Eye was conceived as one such millennium project. It was envisaged as a fun structure for central London, one which people could enjoy by taking a ride to look over one of the world's greatest cities and which would also be a landmark structure in its own right.

The considerable design, manufacturing and construction expertise available on a European-wide or, if necessary, worldwide, basis for projects constructed in the UK meant that the considerable technical difficulties which would have to be surmounted to realise the structure could almost certainly be overcome, give the necessary will. It proved

Figure 17. General view of the London Eye

difficult to find a team ready and committed to carry out the work. There was no prototype from which to work.

The need could be formulated as: To celebrate the millennium with a structure that would delight and provide spectacular views over London.

The Creative Response – the Vision

The concept of an observation wheel was established by the architect, Marks Barfield, in 1994. Initially, there was little interest in the project, but private financial backing was finally secured from British Airways in 1998, and the design, technical, manufacturing and construction issues could then be fully addressed.

Contextual Constraints

- Establish relationships with client team and socially affected groups
- Selection of team with necessary professional competence, knowledge of natural laws

Figure 18. Erection of wheel

- Surveys of context – physical, climatic, topographical, materials, energy
- Legal requirements – planning permissions, legislation.
- Funding and value for investment, competition
- Programme requirements
- Health and safety issues
- **Summary of contextual items**

An outstanding feature of London is the River Thames. The city has grown and prospered around the river, because of the trade and travel opportunities it has provided. In central London the river is the focus of government, religious, administrative, entertainment and other buildings.

The London Eye was to be located close to these other landmarks, alongside the south bank of the river, opposite the Houses of Parliament and close to other centres of entertainment. The intended site was a narrow river bank site with restricted access. On one side was the listed County Hall Building, now used as an hotel with the London Aquarium in its basement. On the other sides were the Jubilee Gardens and the river itself, the latter with a listed, *i.e.* legally protected, embankment. There were a lot of vested interests in the Jubilee Gardens, and ownership of land is very complex. The embankment is a key flood defence, there are environmental issues on the foreshore, and parts of the river bed have to be protected. There was opposition to spoiling the "London skyline" and for the Eye being out of context.

Because of the presence of the County Hall building and the aquarium, restrictions were placed on noise and vibration and on allowable working hours. The Environment Agency placed other restrictions on river bed activity. Queen's Walk is a public thoroughfare with the resultant opposition to closure.

As a facility intended for use by large numbers of the public, out to enjoy a fun ride, exceptional standards of safety were required for the completed structure. The fact that it was a high profile project in a prominent location, using "state-of-the-art" technology was also an added incentive to achieving high standards of safety during construction, it was also a key commercial issue. The sole income from the London Eye Co. comes from the Eye. Losing public confidence could produce disastrous results economically. Because of the key issues of safety, a written Safety Code has been produced. This protects the company against allegations of not looking after the public; considering that the safety of

1500 persons an hour has to be addressed, health and safety legislation is governed by the same requirements as rides in theme parks.

Lambeth Council were uncertain as to how to give planning permission for the project. They gave outline planning permission, and then full permission was given at a special meeting. This was originally restricted to 5 years as a temporary measure.

However, because of the proven value of the Eye to local employment and business and income generation through tourism, it has been given extended planning. It is known that there would now be strong opposition to its removal.

Funding was provided by British Airways on the basis that the Eye would open for the start of the new century (*i.e.* at the beginning of 2000). Because of the delay in securing banking and because of an initial false start to the process of procurement, in early 1998, this gave only 16 months to complete the project. It was a requirement that the Eye should be self-financing so that capital and running costs should be recoverable from revenue. This meant that throughput and a high reliability of operations were essential.

A team of specialists was assembled to work with the architect to develop and deliver the scheme. These specialists were sourced from all over Europe to provide the required expertise in all areas of the project. They were appointed on a design and build basis. A particular problem lies in the fact that the structure is a "machine". There is therefore to a much greater extent than on a standard building a need for the multi-disciplinary skills of the architect, engineers (civil, structural, mechanical, electrical and control), surveyors, planners, *etc.*

Effective Creativity

- Interdisciplinary discussions
- "Brain storming" sessions
- Preparation of alternative imaginative schemes
- Programme for delivery
- **Selection of a preferred solution and its clear definition**

The Eye was the biggest observation wheel in the world when built, providing the highest point in London accessible to the public. It was not only alongside the river, but was cantilevered over it. It is a pleasing and elegant structure in its own right. The first scheme was to be 150 m high with more capsules, but cost savings meant this was reduced to the present 135 m and fewer capsules.

The public are provided with a safe protected view from within enclosed glazed capsules, with minimal obstruction (and hence the capsules are placed on the outside of the rim). A smooth ride with continuous rotation and safe and rapid means of embarkation and disembarkation has been achieved. The continuous rotation was required to achieve the passenger throughput and was a prerequisite of the business case. The Eye has to be self-financing. In practice it was acknowledged that access had to be available for wheelchair users. The operational schedule was therefore adjusted to accommodate some discontinuous sessions from time to time.

The technical demands of the project were met by the appointment of a team of specialists, up to date with the latest design, manufacturing and construction technology. One of the major challenges involved in delivering the project was to ensure the coordination and integration of the work of these specialists.

The demands of the programme and the difficulties of restricted access to the site were to be mitigated by maximising prefabrication and off-site testing and by using the river to transport to site and to lift into place units which were as large as possible. The fabrication of the bulk of the components in the Netherlands meant that they could be delivered by sea as just six pieces (one spindle, one set of legs and four rim sections).

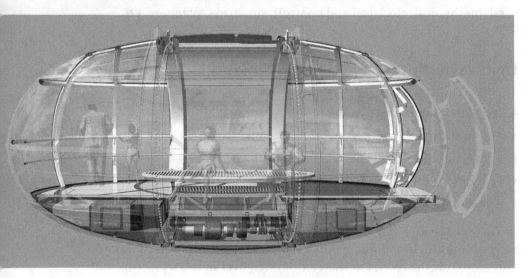

Figure 19. Capsule

▞ Delivery

- Agreement and documentation of preferred solution – continuous reviews of delivery against aims – the design audit
- Definition of management team and process of delivery
- Selection of key designers, of contractors, sub-contractors, and specialists
- Definition of their roles
- Reference to interested third parties
- Agreement on financing – budget, cash flow, cost control procedures
- Agreement on delivery programme
- Agree contract procedures
- Assembly of project/product resources – skilled crafts, materials, equipment
- Agree quality management procedures – design, manufacture, construct
- Health and safety considerations
- Legal requirements and programme
- **Hand over to clients**

The contractual approach initially chosen to deliver the project was to appoint a single contractor who would deliver the whole project. The bidding and negotiation process started on this basis in early 1998. However, it became apparent that this approach was unlikely to be successful, and in the autumn of 1998 it was decided to proceed with a construction manager who would coordinate the work of a number of specialist design and building trade contractors.

The client set up the London Eye Company to deliver the project, and this in turn employed the architect, an independent checking engineer (required by law for public rides) and the construction manager. The latter then prepared the trade contracts, coordinated the work of the trade contractors, secured from a total of six European countries, and drove the programme. A key component of success was the total will to win, no one was going to let the side down.

The wide geographical dispersal of procurement meant that considerable effort went into the coordination of technical issues, as well as the delivery of plant and prefabricated items to site at the right time and in the right sequence.

The building of the Eye was a significant multi-disciplinary project with advanced engineering required in all disciplines and the coordination of structural work with complex electro-mechanical systems. The work varied from the design of the Eye itself for static and dynamic behaviour to the production of unique 3D glass panels for the capsules, the drive system and the boundaries between French, Dutch, English, Italian and Czech trade contractors, involving communication in six languages!

Safety and reliability were paramount. In the light of this, detailed risk assessments were carried out and appropriate systems specified for the degrees of duplication, redundancy and the magnitude of the safety factors which were to be employed. Matters of risk assessment are more appropriate to the concept as a whole and how we recover from various fault scenarios: a fire on board, a capsule motor breakdown, control systems faults. If the Eye were to stick, can rescue be effected? This has direct effects on the configuration of the mechanical systems since it was recognised that a component can break or fail. This had a very significant effect on the configuration of the electronic controls system.

The close involvement of the architect, the independent checking engineer and the client provided useful additional overviews of the whole project and also helped decisions to be taken quickly when necessary. To structural engineers the word "design" is linked very closely with calculations. On a project like the Eye, design is much more the marrying together of many differing components.

The method of construction adopted was to prefabricate the rim in four sections, to transport these by barge up the river and to assemble them into the complete rim and wheel (complete with hub, spindle and supporting legs) by laying all the elements horizontally on temporary platforms, generally piled into the river bed. The complete wheel was then pulled up into place as a single unit. This approach saved vital project time and also minimised the need for working at night during the risky construction phase. The Eye did revolve on December 31, 1999, carrying the design team. It opened to the public on the March 9 after successful completion of the lengthy process of commissioning the electro-mechanical systems.

Performance in Practice

- Reassessment of design brief
- Possible design development
- Social impact assessment
- Client/user satisfaction
- Economic performance
- Maintenance/operational factors
- **Experience to incorporate in future projects**

The design brief/concept as finally developed has been very satisfactory. The Eye has performed well and won many awards. High client satisfaction has been reported.

In one regard the Eye is not performing quite as intended. It was designed to revolve continuously but such are the number of disabled visitors who are now reported to be going on the Eye for a ride that it stops momentarily several times a day to accommodate them. That is perhaps a valuable restriction on the operator's freedom!

In the first year of operation 3.2 million people were carried, against a forecast of 2.2 million. It now carries about 15 million.

The Eye has attracted many visitors to the South Bank of the River Thames in addition to those who attend the nearby entertainment complex containing the Royal Festival Hall, the National Theatre and other venues. It has also provided another landmark structure to be admired in its own right. It has to some degree moved attention from the more prestigious North Bank of the river to the South Bank, making it perhaps even more attractive to visitors.

There are lessons to be learnt about maintenance and operational requirements. The Eye is a machine. As such it has to be maintained on a continuous basis, partly to assure the generation of income and partly to ensure safety.

The philosophy of maintenance is not just to find and fix, but to anticipate and maintain the structure in advance of any problems arising. There are significant implications on good design to achieve rapid and easy maintenance. Although this was considered during development, it is only with experience that the full implications can be seen. For example, there are 32 capsules. Half an hour of maintenance on each means 16 man-hours of work or at least two staff on all night.

There is an annual shutdown, but attempts are being made to reduce the time needed.

Another factor to be considered is the rapidly changing technology. For example, the electronics in the capsule are now obsolete, and some changes are in progress.

One of the biggest lessons learnt is one frequently neglected in the UK. If you want a successful project you have to give the design and construction team freedom and support, and to get them to act as a team committed to winning. Set up a complex project with a fixed budget and severe contractual terms, and you will end up with difficulties. Set up a design without involving the fabricators and builders, and you may end up in difficulties.

How did the team do so well in 16 months? British Airways, the client, told the team what was wanted, paid them to do it, and ensured that they worked together.

4 Case Study 6:
Lesotho Hydro-electric Project

 Formulation of Need

- The cultural societal and physical ethos in which the project is to be carried out
- The history of the project, how did the perception of need arise?
- The decision-making structure – social, client, design management
- The human resources available – professional skills, research, available craft skills
- The physical, technical and economic resources available
- Sustainability requirements, energy sources, environmental impact
- Relevant research and development
- **The clear formulation of the need**

This project provides an interesting example of the methods by which the environmental and social aspects of a major project can and should be included in the project management arrangements of a major programme of construction works.

Project Background

The Lesotho Highlands Region covers one-third of the area of the Kingdom of Lesotho. This small country is completely surrounded by the Republic of South Africa. The region is the source of the Senqu/Orange river, the largest and longest river in the area. The Kingdom of Lesotho is poor and relatively undeveloped. Many parts of the highland region are inaccessible except on foot or horseback.

The primary needs served by the Lesotho Highlands Region project are

- to redirect some of the water which presently flows out of Lesotho northwards towards the population centres of the Republic of South Africa
- to generate hydro-electric power in Lesotho using the redirected flows
- to provide regional social and economic development, water supply and irrigation in Lesotho

 # The Creative Response – the Vision

Contextual Constraints

- Establish relationships with client team and socially affected groups
- Selection of team with necessary professional competence, knowledge of natural laws
- Surveys of context – physical, climatic, topographical, materials, energy
- Legal requirements – planning permissions, legislation.
- Funding and value for investment, competition
- Programme requirements
- Health and safety issues
- **Summary of contextual items**

These objectives are to be achieved by means of the construction of a series of dams, tunnels, pumping stations and hydro-electric works. Lesotho will benefit from the additional development of new roads, telecommunications, health clinics, community-based rural development projects and tourism.

Earnings from the sale of its water will also constitute a major part of Lesotho's export earnings and provide much needed income to aid further development.

The total value of the projects in the Lesotho Highlands Region is in excess of £2 billion. The first phase, with a value of £84 million, opened in 1998.

The possibility of transferring water in this way was first recognised in the 1950s. A number of reports were prepared, culminating in a feasibility study begun in 1982 and completed in 1986. The feasibility study was carried out jointly on behalf of the governments of Lesotho and South Africa.

The scope of the works and its cross-boundary effects involved the formation of a treaty between Lesotho and RSA. This was signed in 1986, and sources of finance were secured with assistance from the World Bank.

Control of Environmental and Social Effects of the Project

It was accepted that such projects bring disadvantages as well as benefits. Local communities would be displaced by inundated land, an influx of foreign workers and job seekers could damage local culture and traditions, together with the negative effects of many other environmental and social impacts. It was recognised, however, that the positive benefits would outweigh the drawbacks, and provisions was made in the treaty stipulating that the standard of living of the affected population should not be inferior to that existing prior to project implementation.

The Lesotho Highlands Development Agency (LHDA) was formed, under the control of the Ministry of National Resources of Lesotho. LHDA acts as the owner of the project, engages engineers and contractors, and co-ordinates the project with government ministries and public corporations.

The structure of the LHDA organisation which manages the project is shown in the accompanying diagram. It can be seen that executives were created to control finance, engineering and construction, operations and environment and public affairs. No specific national environmental guidelines or procedures were in operation in Lesotho prior to the start of this project. LHDA therefore adopted guidelines and detailed environmental specifications from other countries and international agencies, including the World Bank operational directives.

Effective Creativity

- Interdisciplinary discussions
- "Brain storming" sessions
- Preparation of alternative imaginative schemes
- Programme for delivery
- **Selection of a preferred solution and its clear definition**

In order to plan and monitor these requirements, social and environmental considerations were highlighted and given their own place at a high level in the management structure of the project.

These specifications covered aspects within the control of the contractors such as water discharge from the works, protection of flora and fauna, spoil dumps, dust control, landscaping and grassing and general rehabilitation. Further obligations for health and safety, provision of access for local communities to the project health facilities and local community relations were also specified.

LHDA commissioned a series of social and environmental studies and developed an environmental action plan concerning compensation, natural environment and heritage, public health and rural development. Guidelines were drawn up for those communities affected by the project in terms of loss of land, income or resettlement.

 **Delivery**

- Agreement and documentation of preferred solution – continuous reviews of delivery against aims – the design audit
- Definition of management team and process of delivery
- Selection of key designers, of contractors, sub-contractors, and specialists
- Definition of their roles
- Reference to interested third parties
- Agreement on financing – budget, cash flow, cost control procedures
- Agreement on delivery programme
- Agree contract procedures

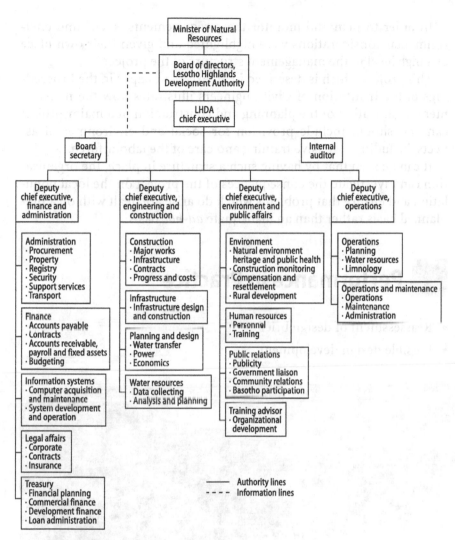

Figure 20. Project organisation for Lesotho hydro-electric project

- Assembly of project/product resources – skilled crafts, materials, equipment
- Agree quality management procedures – design, manufacture, construct
- Health and safety considerations
- Legal requirements and programme
- Hand over to clients

In order to plan and monitor these requirements, social and environmental considerations were highlighted and given their own place at a high level in the management structure of the project.

This project which is described in an excellent paper in the Proceedings of the Institution of Civil Engineers illustrates how the management organisation of the planning and construction of a major project can and should include provision for social and environmental aspects, including extensive training and care of the labour force.

It can be seen that by having such a structure in place, the organisation can investigate the consequences of the project on the local population and ensure that problems which do arise are dealt with on a preplanned basis rather than an inadequate *ad-hoc* basis.

 # Performance in Practice

- Reassessment of design brief
- Possible design development

Figure 21. Project under construction

Figure 22. Completed project

- Social impact assessment
- Client/user satisfaction
- Economic performance
- Maintenance/operational factors
- **Experience to incorporate in future projects**

The scale of the project and the considerable influence upon the whole population of Lesotho required the apparently complex management structure, and the contribution of expertise drawn from many countries proved very effective. A great deal is to be learnt from this project about the organisation and delivery of major works in the developing world.

The projects have made a very major and valuable contribution to the quality of life for both the local population and the recipients of power and water outside the country of Lesotho itself.

4 Case Study 7:
Maribyrnong Footbridge

 Formulation of Need

- The cultural societal and physical ethos in which the project is to be carried out
- The history of the project, how did the perception of need arise?
- The decision-making structure – social, client, design management
- The human resources available – professional skills, research, available craft skills
- The physical, technical and economic resources available
- Sustainability requirements, energy sources, environmental impact
- Relevant research and development
- **The clear formulation of the need**

The City of Essendon (east bank), the City of Sunshine (west bank) and Melbourne Water (river authority) in South East Australia jointly required a footbridge that would connect the bicycle paths and provide access to the various facilities on both sides of the river. The bridge site is located on a flood plain where the Maribyrnong River is, normally, contained within the 64 m width between banks. However, in times of flood, the water has been observed to spread over a 400–500 m width and to rise nearly 2 m above bank level.

It was important not to affect the environment on either side of the river adversely.

The primary initiative was taken by the Essendon Council. The principal consultant, Maunsell Pty Ltd., commissioned BSC Consulting Engineers to carry out the design in 1993. The Council required a low-level type structure in laminated Jarrah timber spanning over

64 m of waterway. The project extended the limits of timber bridging and required research and development.

The need can be summarised as: To provide pedestrian access between the east and west banks of the Maribyrnong River, Melbourne.

 # The Creative Response – the Vision

Contextual Constraints

- Establish relationships with client team and socially affected groups
- Selection of team with necessary professional competence, knowledge of natural laws
- Surveys of context – physical, climatic, topographical, materials, energy
- Legal requirements – planning permissions, legislation.

Figure 23. Site for Maribyrnong footbridge, Melbourne

- Funding and value for investment, competition
- Programme requirements
- Health and safety issues
- **Summary of contextual items**

BSC Consulting Engineers are experienced in the use of timber as a structural material. Jarrah was a timber preferred by the City of Essendon and is a durable Australian hardwood which does not require preservative treatment. It is located in the forests of Western Australia, and its logging is strictly controlled. Other timbers were considered briefly: Douglas Fir and Radiata Pine. These were both imported products whose size would have been significantly greater than the Jarrah beams. The beams were fabricated in Perth and trucked 3200 km from Perth to Melbourne.

The site was thoroughly investigated, and information was available on the physical context of the project. Soil conditions were not ideal for an arched bridge, a 2 m stiff clay crust overlay some 8 m of soft river silt, overlying dense gravels. Flood level was about 1.8 m above bank level, requiring the ends of the bridge to be as high as possible.

Care was taken to maintain existing cycle paths, gardens and wetlands. The design was adjusted to minimise effects on flora and fauna.

Effective Creativity

- Interdisciplinary discussions
- "Brain storming" sessions
- Preparation of alternative imaginative schemes
- Programme for delivery
- **Selection of a preferred solution and its clear definition**

Design decisions involved the timber suppliers and contractors, the local authorities and user groups and the experienced designers.

After careful consideration of the totality of the site constraints and the requirements, it was decided that a clear span of 68 m was required with a low-level structure. This was some 20 m longer than the existing 50 m span of the Greensborough Footbridge, a previous BSC design, known to be the longest of its type at that time.

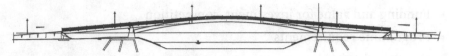

Figure 24. Final design

The two pinned arch consisted of three 1159×120 mm laminated Jarrah sections, 22 m, 23.5 m and 22 m long. At the abutments the arch ends were pin jointed via rigid steel end frames. The deck was 150×30 mm striated Jarrah, forming a non-slip surface. A clear walkway width of 2200 mm was maintained over the entire length of the bridge. A "chunky" hand rail set was adopted to emphasise the use of timber and to increase the damping properties.

Major reviews of design were necessary to maintain the costs within the original estimated budget of $600,000. The final estimated costs were reduced from $850,000 to $690,000. Changes were also necessary in the foundation design to reduce anticipated movements under live load.

A linear elastic analysis was used to set up both the vertical and lateral computer models. Full details are given in the articles listed in the Appendix. Due to the complexity of the model, it was considered necessary to use both 2D and 3D models for design. The 3D model was used initially to check the 2D results and secondly to perform the dynamic analysis. Full attention was also given in the dynamic analysis to pedestrian-induced vibrations and to wind loading.

The final design had massive end abutments on 27-m driven piles into the gravel.

 Delivery

- Agreement and documentation of preferred solution – continuous reviews of delivery against aims – the design audit
- Definition of management team and process of delivery
- Selection of key designers, of contractors, sub-contractors, and specialists
- Definition of their roles
- Reference to interested third parties
- Agreement on financing – budget, cash flow, cost control procedures
- Agreement on delivery programme

- Agree contract procedures
- Assembly of project/product resources – skilled crafts, materials, equipment
- Agree quality management procedures – design, manufacture, construct
- Health and safety considerations
- Legal requirements and programme
- **Hand over to clients**

The major supplier of Jarrah (Bunnings Forest Products Pty Ltd.) was the only organisation in Australia capable of glue laminating the large size sections in Jarrah.

There were problems with the contracting organisations. The first main contractor did not meet his commitments and absconded, stopping work for several weeks. The concrete subcontractor was employed directly to complete the abutments, and a new contractor, Cowling Building Services (CBS), was appointed to complete the timber fabrication. They commenced work in September 1994.

The fourteen glue-laminated beams up to 1159×120 mm were fabricated by Bunnings Laminated Products in Perth. These components, together with associated laminated diaphragms, handrails, balusters and decking, were delivered from Perth to Melbourne and sorted and assembled in a disused factory in a suburb of Melbourne for a trial erection in November 1994.

One off-site trial erection procedure did not work, the heavy sections fell, smashing most of the timber members and handrails. Five new beam sections were fabricated by Bunnings and supplied in January 1995.

A carefully considered erection plan was prepared, using piled platforms in the river at the ends of the central bridge section. Piling was halted when the pile driver hit sunken car bodies. Despite these setbacks CBS persisted, enabling the erection of the bridge on April 24, 1995.

On May 9, 1995, the temporary supports were slowly lowered until the bridge was self-supporting. The centre span dropped in height by some 45 mm, as predicted.

The bridge was officially opened by the Australian Federal Treasurer, the Hon. Ralph Willis, on July 2, 1995.

Figure 25. Completed project

 # Performance in Practice

- Reassessment of design brief
- Possible design development
- Social impact assessment
- Client/user satisfaction
- Economic performance
- Maintenance/operational factors
- **Experience to incorporate in future projects**

In the first 12 months the bridge was used by thousands of pedestrians and joggers.

It has served its purpose well. Many lessons have been learned from this innovative project, which will no doubt be of great help in future projects.

Mr. Carlin-Smith, managing director of BSC, stated that, "The bridge stands as a monument to the integrity and perseverance in the face of adversity of all those involved in its design and construction".

Case Study 8:

Bahá'í Temple – Delhi

 Formulation of Need

- The cultural societal and physical ethos in which the project is to be carried out
- The history of the project, how did the perception of need arise?
- The decision-making structure – social, client, design management
- The human resources available – professional skills, research, available craft skills
- The physical, technical and economic resources available
- Sustainability requirements, energy sources, environmental impact
- Relevant research and development
- **The clear formulation of the need**

The Bahá'í faith, built on the teachings of Bahá'u'lláh in the nineteenth century, embraces the belief in the unity of God and his prophets. It promotes harmony through religion hand in hand with science. The construction of Houses of Worship as collective centres for men's souls manifesting the faith and revealing the simplicity, clarity and freshness of the new revelation has been a major part of the philosophy. These buildings are conceived as providing tranquil places of assembly, for meditation and prayer and for readings from holy books for the followers of all religions. Six such temples were built, in the USA, Panama, Kampala, Frankfurt, Sydney and Western Samoa. The need was perceived for a seventh, on a remaining continent, and this, the mother temple of the subcontinent of India, is a beautiful example of the integration of architecture with structure. It is situated at Bahapur, on the outskirts of New Delhi. It has a main hall accommodating 1200 seats leading to the entrance halls on a main podium for use as places for prayer, foyers and circulation. This House of Worship is surrounded by pools. An ancillary building built into the surrounding pools provides rooms for a library and general reception and administration purposes.

Fariborz Sahba, the Bahá'í architect and the project manager selected, had achieved eminence in design of beautiful buildings in Iran. He arranged for the appointment of the Flint and Neill Partnership as structural engineers, who had designed the structure of the Bahá'í temple in Panama and were currently the consultants for the temple in Western Samoa. They had particular experience in design of concrete shells and in seismic and wind engineering. They subsequently obtained the assistance of Mahendra Raj Consultants of Delhi in advising on local design practices and materials and in supervision of the works. In the course of the design, advice on concrete materials was provided by the Cement & Concrete Association, aggregate testing was undertaken by Messrs. Sandberg, and wind tunnel testing of a model to derive wind loading and natural ventilation characteristics was undertaken in the Aeronautical Department of Imperial College, London.

Figure 26. Overall site plan/sections for Bahá'í temple, Delhi

In the early stages of the structural design, the engineers and architects visited Delhi to explore the capability of Indian contractors who might be chosen to tender for the construction contract, to investigate sources of suitable materials and to locate local engineers who might provide guidance on relevant Indian practices. These visits confirmed that the necessary professional skills could be mobilised in the sub-continent.

The need was to provide a temple in the Indian sub-continent similar to the six temples already developed elsewhere.

 # The Creative Response – the Vision

Contextual Constraints

- Establish relationships with client team and socially affected groups
- Selection of team with necessary professional competence, knowledge of natural laws
- Surveys of context – physical, climatic, topographical, materials, energy
- Legal requirements – planning permissions, legislation.
- Funding and value for investment, competition
- Programme requirements
- Health and safety issues
- **Summary of contextual items**

As the project manager the architect provided the liaison between the Bahá'í World centre, its Indian representative: the National Spiritual Assembly of the Bahá'ís of India as the client, and the consultants.

The temple was to be located on a site in the capital of India, on the outskirts of New Delhi. This is in a zone of moderate seismic activity subject to high summer temperatures and atmospheric pollution from neighbouring industries. The site is underlain by heavily fissured quartzite with randomly orientated stratification containing large lenses of mica beneath decomposed rock and soil, and this led to the decision to adopt concrete pad foundations.

Funding came from the Bahá'í Universal House of Justice using voluntary contributions from members of the faith worldwide. Competitive tenders for the construction were obtained from Indian contractors.

There were no specific constraints on timetable, although the construction contract had a completion time of 2.5 years. There was no period specified for completion of the preliminary investigations, design and preparation of contract documents. For various reasons the construction actually took 6 years and 8 months, the contract being re-negotiated after an interval when the work was stopped before construction of the shells.

Normal health and safety procedures were observed.

Effective Creativity

- Interdisciplinary discussions
- "Brain storming" sessions
- Preparation of alternative imaginative schemes
- Programme for delivery
- **Selection of a preferred solution and its clear definition**

A joint office for the architect and the structural engineer was opened in London in which the structural design was developed from the architect's concept. For this project the goal was the attainment of the concept, and no alternative schemes were on the agenda. Initially, the selection of the preferred solution hinged on the choice of alternative compatible geometric forms suitable for construction and providing adequate structural strength, and this was achieved by close collaboration within the team. "Brain-storming" sessions were held in the joint office of the engineers and architect.

The design for the temple as conceived by the architect, Fariburz Sahba, takes the form of an opening lotus flower, symbolic of the culture of India and representing the Bahá'í vision. The opening lotus not only has an association with all Indian religions but is also one of the most beautiful flowers in the world. It grows in swamps and raises its head out of the slime absolutely clean and perfect. The Bahá'ís perceive the temple as "a fragile flower enshrining an idea, the idea of light and growth, a shelter of petals interposed between earth and sky".

It was decided to construct the petals as pure shells of reinforced concrete, retaining simplicity of concept as well as weather tightness and obviating the need for internal finishes. Finite element analyses were undertaken using non-linear deflection predictions. No movement joints in the roof structure were found necessary due to the relative insensitivity of the stresses to temperature effects. Foundations were concrete pads. Continuity of the arch rings was an essential feature of the design and prohibited the provision of movement joints in the ring or in the adjacent podium slab. The outer supports to the podium between the pools are provided with neoprene pad bearings to accommodate thermal and shrinkage movements. The foundations to the primary columns beneath the junctions of the entrance and outer leaves were also mounted on rubber pads. A further neoprene pad was provided above the interior dome hub, supporting the radial roof beams.

The profiles of the outer surfaces of the roof shells evolved from the architect's conceptual drawings and model, and were constrained by his target shapes and dimensions. Geometric forms considered to be potentially feasible were those which could be explicitly defined by mathematical equations and which would permit the use of repetitive patterns of shuttering to concrete and of external finishes or cladding. The shapes were developed by a process of repetitive selection of trial forms and their basic setting-out points. The detailed dimensions of each of the marble cladding panels were later calculated by the engineers.

A system of Cartesian coordinates for every segment of the temple was computed from which levels and distances used to set out the surfaces and boundaries were derived. Eighteen reference stations were established outside the building for setting out the arches, entrance, outer and inner leaves. The development of the design relied heavily on the use of computing technology not available a few years earlier.

Feasible mathematical forms for the shells were calculated by the engineers and visually displayed to enable the architect to view them from any point.

Stress analysis was undertaken for self-weight and superimposed loads, seismic response, wind loads and overall and differential temperature, using programs developed by the engineers. Computer analysis of the temporary staging was subsequently undertaken for the contractor at the Civil Engineering Department of the Indian Institute of Technology, Madras.

The structural design was generally in accordance with Indian standards using data provided by the Delhi meteorological office. The engineer and contractor established detailed method statements, acceptance criteria and checklists based on the intended sequence of construction, the assumptions made in the design of the formwork, the procedures developed from mock-ups and the tests carried out on materials. A concrete laboratory on site carried out mix designs for various grades of concrete and exercised strict control on concrete quality.

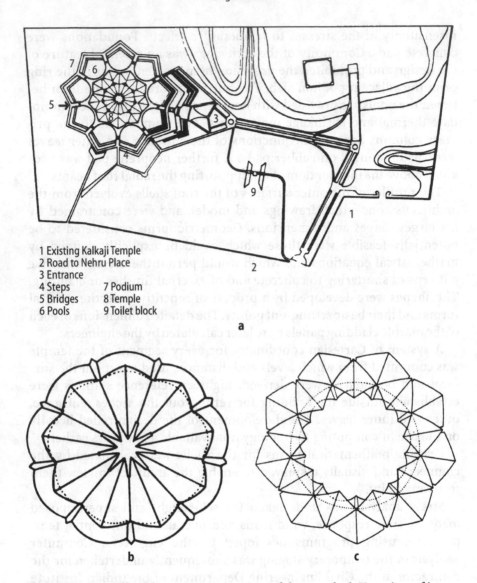

1 Existing Kalkaji Temple
2 Road to Nehru Place
3 Entrance
4 Steps   7 Podium
5 Bridges   8 Temple
6 Pools   9 Toilet block

a

b

c

Figure 27. View of building on site. **a** Layout of the temple. **b** Top view of inner leaves and arches. **c** Top view of entrance and outer leaves.

Delivery

- Agreement and documentation of preferred solution – continuous reviews of delivery against aims – the design audit
- Definition of management team and process of delivery
- Selection of key designers, of contractors, sub-contractors, and specialists
- Definition of their roles
- Reference to interested third parties
- Agreement on financing – budget, cash flow, cost control procedures
- Agreement on delivery programme
- Agree contract procedures
- Assembly of project/product resources – skilled crafts, materials, equipment
- Agree quality management procedures – design, manufacture, construct

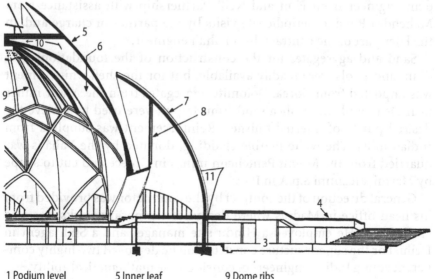

1 Podium level	5 Inner leaf	9 Dome ribs
2 Pool deck level	6 Interior dome	10 Crown of dome
3 Pool	7 Outer leaf	11 Entrance leaf
4 Bridge	8 Glazing	

Figure 28. Plan and elevation – section through inner and outer leaves

- Health and safety considerations
- Legal requirements and programme
- **Hand over to clients**

The detailed structural drawings, specifications and bills of quantities were prepared by the structural engineer. The computer-based drawings and structural analysis were checked by Dr. H. Naimi of Geneva. The design of all temporary works was undertaken by the superstructure contractors. A number of mock-ups of formwork were constructed which provided a further check on geometry.

The engineer and contractor established detailed method statements, acceptance criteria and checklists based on the intended sequence of construction, the assumptions made in the design of the formwork, the procedures developed from mock-ups and the tests carried out on materials. A concrete laboratory on site carried out mix designs for various grades of concrete and exercised strict control on concrete quality.

The sub-structure contract was managed and supervised by the engineers. For the contract for the superstructure, the architect in person acted as the client's project manager on site in charge of budget control, public relations and supervision of all architectural work and of appearance of materials and finished structural work. Supervision of the structural works on behalf of the client was undertaken by a resident engineer from Flint and Neill Partnership with assistance from Mahendra Raj and periodic site visits by the partner in charge and, in the later part of the contract, by a Bahá'í engineer.

Sand and aggregates for the construction of the foundations, podium and pools were readily available, but for the shell white cement was imported from Korea. Dolomite aggregates from the Alwar mines near Delhi and white silica sand from Jaipur were used to achieve the desired purity of internal finishes. Reinforcement was supplied from Indian mills. The white marble cladding, donated by the Bahá'ís was quarried from the Mount Pentelicon mines in Greece and cut to shape by Marmi Vicentini S.p.A in Italy.

General direction of the contract by the contractor was provided from his head office in Madras where design of temporary works was also carried out. He employed a Hindu site manager and a Sikh agent in Delhi. For the superstructure construction he deployed two highly competent young Indian engineers of considerable mathematical ability.

A preliminary contract for the excavation of the foundations was let to Ahlwalia Construction Co. The construction contract was let to ECC Construction Group of Larsen & Toubro Ltd, Madras, on the basis of tender documents prepared by the engineers in collaboration with the architect.

Design and production of the concrete mixes employed state of the art methodology including control of concrete temperature during placing in hot weather. The construction of the formwork to the shell required very tight tolerances, and the provision of bush-hammered internal finishes with architectural patterns entailed highly complex design of staging and the building of mock-ups. The contractor employed highly skilled Indian carpenters with specialists to form the fibre-reinforced linings for the inner leaves.

Unusually stringent requirements on the internal appearance and the decision to cast the shells *in situ* led to the need to suspend all reinforcement from outer shutters, and to design and erect formwork to enable continuous concrete placement in lifts up to 14 m in height.

Some of the construction technologies adopted were not the most sophisticated, and in some cases ancient traditional methods were deliberately resorted to. For example, much of the concrete for the shells was transported in bowls from the batching plant by women. The hand pick-hammered concrete finishes far surpassed the quality achievable by mechanical means.

 # Performance in Practice

- Reassessment of design brief
- Possible design development
- Social impact assessment
- Client/user satisfaction
- Economic performance
- Maintenance/operational factors
- **Experience to incorporate in future projects**

The apparently complex design and construction processes have resulted in a restful and elegant building. The restful nature of the completed building is very apparent as soon as the site is entered, contrasting with the very active and bustling nature of Delhi.

Formulation of Need

- The cultural societal and physical ethos in which the project is to be carried out
- The history of the project, how did the perception of need arise?
- The decision-making structure – social, client, design management
- The human resources available – professional skills, research, available craft skills
- The physical, technical and economic resources available
- Sustainability requirements, energy sources, environmental impact
- Relevant research and development
- **The clear formulation of the need**

Hampden Gurney School replaced a two-storey school building dating from the 1950s that occupied a site bombed in World War II. As a Church of England School that prides itself on its high academic standards, Hampden Gurney had sought for a number of years to provide its pupils with a school worthy of the twenty-first century.

The need was to provide a modern primary school as the centre piece of a recreated Marylebone city block overlooking the constant activity of Edgware Road, London.

The Creative Response – the Vision

Contextual Constraints

- Establish relationships with client team and socially affected groups
- Selection of team with necessary professional competence, knowledge of natural laws
- Surveys of context – physical, climatic, topographical, materials, energy
- Legal requirements – planning permissions, legislation.
- Funding and value for investment, competition
- Programme requirements
- Health and safety issues
- **Summary of contextual items**

Procurement was by a design, develop and build process, with the school trustees acting as landowners and user clients. They chose the

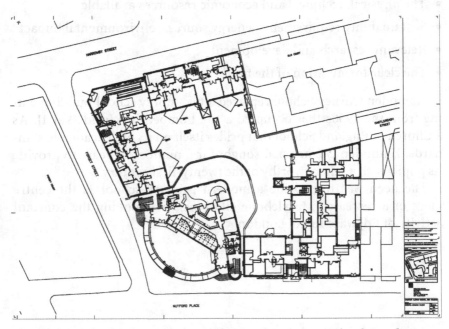

Figure 29. General arrangement of Hampden Gurney School, London

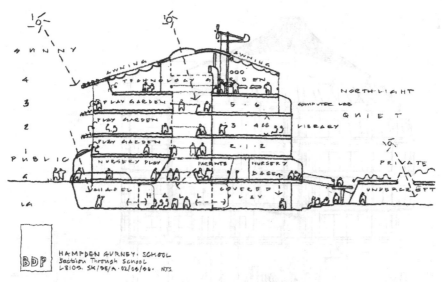

Figure 30. Sketch of cross-section

architects and design team (Building Design Patrtnership) in competitive interview in 1995.

The site is a very urban central London area, just north of Marble Arch. The decision to redevelop the bombed area was delayed for many years, with temporary buildings being placed on the site, including a junior school. The increasing need for affordable residential accommodation and for better educational facilities led to the decision to develop the site, and to use the income from the estate to help support the school.

Effective Creativity

- Interdisciplinary discussions
- "Brain storming" sessions
- Preparation of alternative imaginative schemes
- Programme for delivery
- **Selection of a preferred solution and its clear definition**

The multi-disciplinary team was responsible not only for the design of the school but of the two residential blocks also.

The concept was chosen in competitive interview by the head of the school trustees, Father Michael Burgess, who, with his staff, recognised the opportunities of the "vertical school" or "children's tower". The

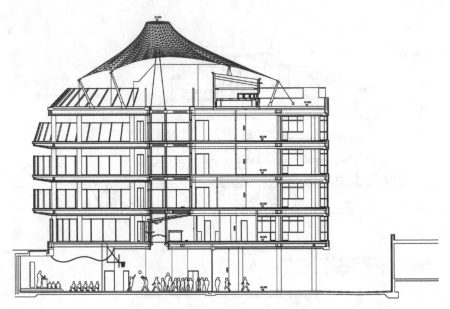

Figure 31. Completed cross-section

open air play decks provide safe weatherproof play space and territory for different age-groups adjacent to their classrooms; they can also be used as open-air classrooms. The building is located on the prominent south-west corner of the site, forming with two new residential buildings a garden courtyard in the innermost part of the site. This gives the school the best aspect for sunlight and a prominence within its neighbourhood. The outward-looking building recognises the trustees aspirations for it to play an active role in the life of the community. It comprises six levels, giving it a presence amongst neighbouring buildings and giving the staff and pupils a new prospect over their location.

Together with the library and multimedia room, the classrooms themselves are set on three levels above ground floor, with a group teaching room on the roof. All teaching rooms enjoy good north light and have a prospect over the courtyard garden formed by the residential buildings. As they advance in the school, the children move up the building, from the nursery on the ground floor to years five and six on the third level. The hall, chapel and music and drama room are set at the lower ground level (one level below the pavement), and a new nursery for 30 children is located at ground level (900 mm above pavement level). Two long ramps and steps give easy access to the entrances for small children and parents with prams. Play areas are located at each teaching level, separated from the classrooms by a bridge across the central lightwell. The play areas are open to the air, their long south sides curved to enjoy all-day sunpaths. The 1.9 m balustrade is formed of Planar glazing uprights. The lower ground level play area is dedicated to ball games and team sports.

The main structure comprises a steel frame crowned with an arched truss at the fourth floor level, high tensile Macolloy bars support the bridge steels in the lightwell, transferring the loads to the truss over-head and allowing the hall to be column free. The outer wall is in a brick sympathetic to the surrounding stock buildings. The roof springs from the steel truss, protecting the light well and creating threshold spaces in the roof play area.

Two factors governed the engineering design: column free space for the play area at lower ground floor, and speed of erection. This led to a steel frame with light weight composite floors, laterally restrained by two concrete staircases.

Figure 32. Photograph of school

The column-free 16 m span above the central atrium is achieved by hanging all floors off the structure at mid-span. A bow arch at roof level picks up the loads by means of a fan of cables. Macalloy bars transfer the tension load from floor to floor.

During construction temporary columns were used in compression to support the structure. Once the top level was reached and the arch truss built, the fanning cables and high tensile rods were tensioned and the compression columns removed. The Macalloy bars were designed so that the failure of a bar or pair of bars would not cause the structure to collapse. The structure was also designed against a minimum requirement for vibration to provide comfort to the users. A value engineering exercise demonstrated that the system was more efficient and more economical than traditional transfer structures such as plate girders, while also providing the school with a landmark architectural feature.

Delivery

- Agreement and documentation of preferred solution – continuous reviews of delivery against aims – the design audit
- Definition of management team and process of delivery
- Selection of key designers, of contractors, sub-contractors, and specialists
- Definition of their roles
- Reference to interested third parties
- Agreement on financing – budget, cash flow, cost control procedures
- Agreement on delivery programme
- Agree contract procedures
- Assembly of project/product resources – skilled crafts, materials, equipment
- Agree quality management procedures – design, manufacture, construct
- Health and safety considerations
- Legal requirements and programme
- **Hand over to clients**

After a prolonged period, Westminster City Council Planning Department approved the redevelopment of the site to restore the pre-war urban grain, and supported the idea of a new school and housing complex. The Department of Education and the local education authority recognised the unusual design and exercised their discretion, particularly in requirements for external play, given the school's urban location. Planning approval was given in October 1998.

Construction began in July 2000, and the £6.0 million building was completed for the start of the spring term in January 2002.

The project had to be phased to facilitate the operation of the existing school, with the new school and south housing block being completed as phase one. As well as the 3,344 m^2 school, which provides for 240 children aged from three to eleven, the £18 million development includes a 6,000 m^2 residential development of 52 apartments in two six-storey blocks, the profits from which funded the redevelopment of the school. The school retains the freehold to the entire site, giving an assured ground rent income.

 # Performance in Practice

- Reassessment of design brief
- Possible design development
- Social impact assessment
- Client/user satisfaction
- Economic performance
- Maintenance/operational factors
- **Experience to incorporate in future projects**

As a radical alternative to the standard storey school, Hampden Gurney stands out as a bold answer to a common inner-city problem. As an infill stitching together a long fragmented urban block and providing much needed housing in central London, it is also an excellent demonstration composite development. The whole package was financed by the residential development.

Case Study 10:

Bridge Academy

 Formulation of Need

- The cultural societal and physical ethos in which the project is to be carried out
- The history of the project, how did the perception of need arise?
- The decision-making structure – social, client, design management
- The human resources available – professional skills, research, available craft skills
- The physical, technical and economic resources available
- Sustainability requirements, energy sources, environmental impact
- Relevant research and development
- **The clear formulation of the need**

From the 1970s to about 2000 there has been limited investment in education in the UK. The new Labour government of 1997 placed education high on the agenda with the famous slogan "Education, Education, Education". Since then, there have been a number of initiatives in primary, secondary, further and higher education. The Academy programme has been one of those initiatives. The programme was developed to replace failing schools with new facilities. The intention is that the new building facilities would attract motivated staff and encourage an increased sense of achievement and responsibility in the students. This is intended to encourage older pupils to continue their education into the higher or further educational concepts, as a form of post-school training for those not taking A level, university or college training and qualifications.

The financing of the new projects is primarily through government, but approximately 10% of the costs are from the private sector. The private sector portion of the costs allows sponsors to oversee the detail of the development and is also considered as a way to attract private sector know-how into the education system. Each Academy has a specialism in which it provides enhanced learning. The private sponsor for the Bridge Academy is the UBS bank, and its specialist subject matters are mathematics and music.

The summary need was: To provide educational facilities on a restricted inner city site, focusing on mathematics and music specialities, which will be attractive to all students and staff, enabling their development as contributors to society, and supported by private sector interest and finance.

The Creative Response – the Vision

Contextual Constraints

- Establish relationships with client team and socially affected groups
- Selection of team with necessary professional competence, knowledge of natural laws
- Surveys of context – physical, climatic, topographical, materials, energy
- Legal requirements – planning permissions, legislation.
- Funding and value for investment, competition
- Programme requirements
- Health and safety issues
- **Summary of contextual items**

The site is a disused gas works with hydrocarbons, lead, sulphate and semi-volatile organic compound contamination. The height of the building is limited to 20 m to meet local planning constraints. The programme brief on the constricted inner-city site defined generates a multi-storey building varying from six stories to one, with basements up to 6 m deep. The north of the site is bounded by the Grand Union Canal, and access for maintenance must be maintained.

Figure 33. Section of model for Bridge Academy, London

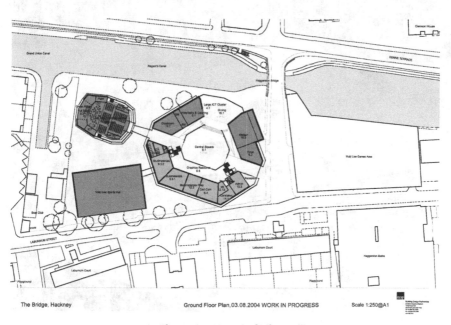

Figure 34. Ground plan

121

Research into the learning process has established the importance of natural light and adequate ventilation, to create stimulating learning environments for the students. To achieve these aims, a closely collaborative multi-disciplinary team of designers is essential. The multi-disciplinary design practice of the Building Design Partnership was ideally suited to this purpose.

Effective Creativity

- Interdisciplinary discussions
- "Brain storming" sessions
- Preparation of alternative imaginative schemes
- Programme for delivery
- **Selection of a preferred solution and its clear definition**

Figure 35. Elevation of model

At an early stage in the design, the structural engineers and architects developed different concepts, with "hoop" structures, surrounding the top of the building. The intention was to create a structure that was in itself stimulating and original. The hoop was to suspend the floors beneath, providing a column-free open space below. From the hoop, inclined members would support the ETFE cladding (ethylene tetra-flouro-ethylene) cover to the central space. These initial concepts had to be reconsidered as the spans and loads were such that beam depths transformed the hoop into 4 m deep trusses.

Two other solutions were developed for consideration by the team:

- The first involved the construction of a conventional building with inclined members supporting the ETFE cladding of the atrium. In this option the hoop as a supporting structure was abandoned.
- The second was similar to the original hoop concept, but reduced the suspension loads and spans; the lower, internal floors are suspended from the higher surrounding structure. The inclined faces of the ETFE-covered walls form the depth of the suspension ring. The hoop at the top of the building forms the top chord of the triangulated system.

The design team agreed that the second option was of greater interest, and provided an opportunity for the use of the visible structure as an attractive feature of the school, and this solution was adopted.

The elements above have resulted in a building circular in plan, around lower, central social spaces that include a library, central social/assembly hall space and dining area, topped with external play terraces. Considering the spans, the loads, the configuration of the building, and the potential congestion of the central London site, a steel-framed structure has been adopted.

 # Delivery

- Agreement and documentation of preferred solution – continuous reviews of delivery against aims – the design audit
- Definition of management team and process of delivery
- Selection of key designers, of contractors, sub-contractors, and specialists

Figure 36. Complete model of superstructure

- Definition of their roles
- Reference to interested third parties
- Agreement on financing – budget, cash flow, cost control procedures
- Agreement on delivery programme
- Agree contract procedures
- Assembly of project/product resources – skilled crafts, materials, equipment
- Agree quality management procedures – design, manufacture, construct
- Health and safety considerations
- Legal requirements and programme
- **Hand over to clients**

As the geometry of the building is circular in plan and sloping in two orthogonal planes, the setting out of the structure and the building was complex. It was decided that the structural engineers would set out the building from a 3D model, and the design team used the model to develop the design. The process is known as building information modelling (BIM); plans are not drawn in traditional two dimensions but are generated from slicing a single 3D model.

This school has not yet been completed, but it is intended to have the construction work completed and to be open for students by September 2007.

4 Case Study 11:

Harris Manchester College, Oxford

 Formulation of Need

- The cultural societal and physical ethos in which the project is to be carried out
- The history of the project, how did the perception of need arise?
- The decision-making structure – social, client, design management
- The human resources available – professional skills, research, available craft skills
- The physical, technical and economic resources available
- Sustainability requirements, energy sources, environmental impact
- Relevant research and development
- **The clear formulation of the need**

Harris Manchester College needed additional accommodation for students and staff. It was about to become a full college of the University of Oxford, after about 300 years of existence as an independent non-conformist institution founded because, until the late nineteenth century, students could not register as students of Oxford University unless they were practicing members of the Anglican community.

The College had no experience of designing or procuring new buildings. The project was to be funded by the Farmington Trust, and I was invited by one of their trustees to advise them on the processes involved. The opportunity reflected my interest in the role of buildings in influencing the quality of thought, feelings and work of we who experience them. In particular, I had been struck by a chance remark I heard in a lecture in Bruges – that throughout Europe and indeed beyond, it was not possible to enter a human settlement without finding some

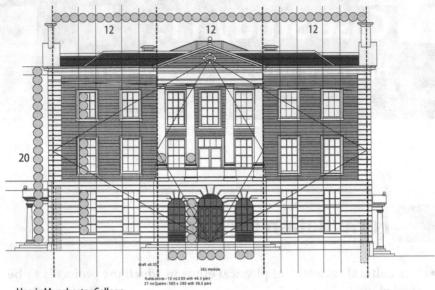

Harris Manchester College
Principal's Lodgings
South Elevation - Geometric Proportions

Figure 37. Principal's residence, proportions; Harris Manchester College, Oxford

evidence of our Christian culture and tradition. Knowing that many churches and cathedrals, as well as works of art, had been deliberately planned to embody certain principles of order and harmony, I felt that it might be of interest and value to explore the relationship between concepts of natural law and the conscious use of order to produce a state of rest and heightened awareness in great buildings. I was invited to appoint a team of designers to carry out the work, and asked to act as chairman of the building committee.

Most of the aspects of defining the need were therefore in place. The architect Peter Yiangou was very sympathetic towards the concepts of elegance, restfulness and simplicity, and the other consultants on structural and services engineering and the quantity surveyor were appointed from our experience, making a very coherent team.

We were able to assess the other contextual factors (physical, economic, planning, *etc.*) and form a clear vision of the need: to provide suitable facilities for scholarly students in inspiring residences.

**Malvern architectural society
Manchester college lecture**

Principal's residence - elevation

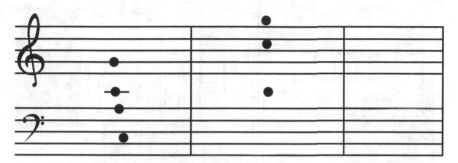

Overall proportions	Portico subdivisions
36	3
24	2
20	1
12	

Figure 38. Principal's residence, harmonies

 # The Creative Response – the Vision

Contextual Constraints

- Establish relationships with client team and socially affected groups
- Selection of team with necessary professional competence, knowledge of natural laws

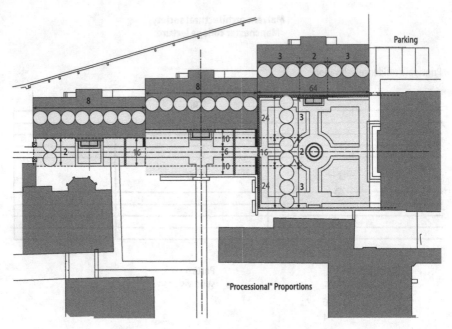

Figure 39. Processional layout, proportions

- Surveys of context – physical, climatic, topographical, materials, energy
- Legal requirements – planning permissions, legislation.
- Funding and value for investment, competition
- Programme requirements
- Health and safety issues
- **Summary of contextual items**

The contextual constraints were clearly recognised during the formulation of the need. We established the necessary legal appointments of designers, and agreed the processes for budgetary and quality control. Physical surveys of the clearly defined site were undertaken, and permission obtained from the city planning authority. We also developed a close liaison with the overall University authorities. Members of other colleges, including All Souls, were included on the building committee.

A formal programme and specification of the accommodation was drawn up and agreed by the client.

**Malvern architectural society
Manchester college lecture**

Processional way - plan

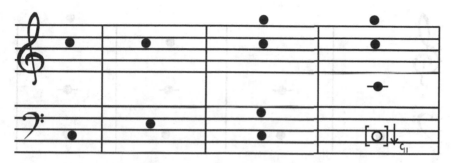

Section A	Section B	Section C	Overall
4	16	8	24
1	5	3	16
		2	8
			2

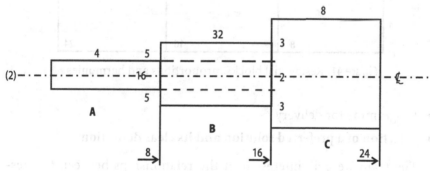

Figure 40. Processional layout, harmonies

Effective Creativity

- Interdisciplinary discussions
- "Brain storming" sessions
- Preparation of alternative imaginative schemes

Malvern architectural society
Manchester college lecture

Three residential blocks - elevation

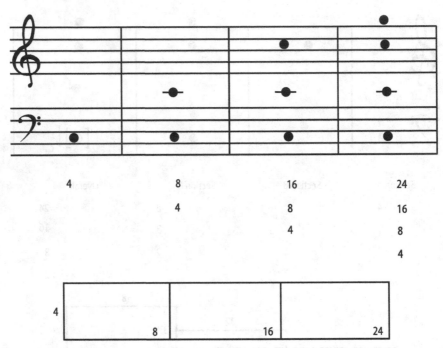

Figure 41. Residential blocks, proportions and harmonies

- Programme for delivery
- **Selection of a preferred solution and its clear definition**

The team were all interested in the relationships between the aesthetic aspects of the design and the effective usefulness of the accommodation. At the time of the Renaissance, a major shift in theological emphasis took place. The Humanist concern with the presence of God-in-Man, as distinct from God-on-High, contributed to the design of many smaller churches, as well as to the detailing of many larger cathedrals. In particular, the Renaissance designers made great use of musical harmonies in their work. This is demonstrated in a memorandum describing the design and layout of the Church of San Francesco de la Vigna in Venice, by Francesco Giorgi. The proportions of the Church can be described musically.

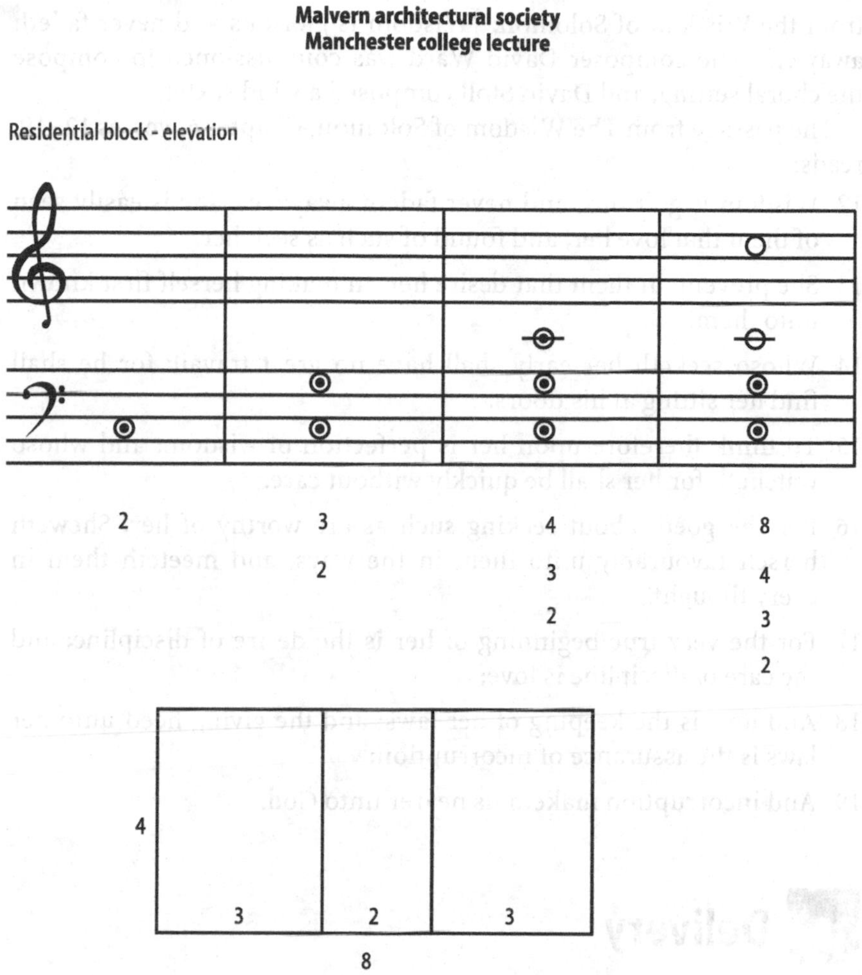

**Malvern architectural society
Manchester college lecture**

Residential block - elevation

Figure 42. Single block, proportions and harmonies

During the formulation of the project design for the residences at the College, it was proposed that we might consider some of the Renaissance techniques of design linking music and architecture, ideas developed by the architect Alberti in his classic treatise on architecture, *The Ten Books on Architecture*. The final agreed design therefore was "fine tuned" to reflect some of the basic musical harmonies in the classical octave. This scheme design was adopted and the preparation of contract documents proceeded. Diagrams are attached giving the basic proportions and their equivalent musical harmonies (Figures 37–42).

The major contributor to the funding of the project then suggested that we use the harmonies to compose a choral setting for a passage from the Wisdom of Solomon: "Wisdom is glorious and never fadeth away ..." The composer David Ward was commissioned to compose the choral setting, and David Stoll composed a wind sextet.

The passage from The Wisdom of Solomon, Chapter 6, verses 12–19, reads:

12 Wisdom is glorious, and never fadeth away: yea, she is easily seen of them that love her, and found of such as seek her.

13 She preventeth them that desire her, in making herself first known unto them.

14 Whoso seeketh her early shall have no great travail: for he shall find her sitting at his doors.

15 To think therefore upon her is perfection of wisdom: and whoso watcheth for her shall be quickly without care.

16 For she goeth about seeking such as are worthy of her, Sheweth herself favourably unto them in the ways, and meeteth them in every thought.

17 For the very true beginning of her is the desire of discipline; and the care of discipline is love;

18 And love is the keeping of her laws; and the giving heed unto her laws is the assurance of incorruption;

19 And incorruption maketh us nearer unto God.

Delivery

- Agreement and documentation of preferred solution – continuous reviews of delivery against aims – the design audit
- Definition of management team and process of delivery
- Selection of key designers, of contractors, sub-contractors, and specialists
- Definition of their roles
- Reference to interested third parties
- Agreement on financing – budget, cash flow, cost control procedures
- Agreement on delivery programme
- Agree contract procedures

Figure 43. Completed Principal's residence

- Assembly of project/product resources – skilled crafts, materials, equipment
- Agree quality management procedures – design, manufacture, construct
- Health and safety considerations
- Legal requirements and programme
- **Hand over to clients**

The delivery process for this project was very conventional. Contract documents were prepared and tenders invited from a short list of carefully selected contractors.

Regular reports were made by the design team to meetings of the building committee, and work went according to plan. During the course of construction one of the sub-contractors went into liquidation, but this problem was resolved without serious disruption of the programme or budget. The completed buildings are very satisfactory and much appreciated by the college management and by the residents. All the factors listed above were considered in the delivery process.

 # Performance in Practice

- Reassessment of design brief
- Possible design development
- Social impact assessment
- Client/user satisfaction
- Economic performance
- Maintenance/operational factors
- **Experience to incorporate in future projects**

The buildings completed were in accordance with the agreed sketch designs. At the end of the project as described above, a lecture was given at the College describing the design process and illustrating the relationship between the musical harmonies and the proportions.

4 Case Study 12:

Angel of the North

 Formulation of Need

- The cultural societal and physical ethos in which the project is to be carried out
- The history of the project, how did the perception of need arise?
- The decision-making structure – social, client, design management
- The human resources available – professional skills, research, available craft skills
- The physical, technical and economic resources available
- Sustainability requirements, energy sources, environmental impact
- Relevant research and development
- **The clear formulation of the need**

In 1986 the Gateshead Borough Council, where there was some feeling of being dwarfed by the City of Newcastle, decided to create a better image of the town with a pioneering scheme called "Art in Public Places": over 20 sculptures and murals for streets, parks, metro-stations and hospitals were developed.

A highlight is the Angel of the North by sculptor Anthony Gormley, winner of the 1994 Turner Prize. This dominating statue now stands on a low hill where two dual carriageways meet. Realising the project took several years. In the early 1990s the Council inherited an old colliery site on the town's southern boundary. The concept of a landmark sculpture was developed. The need could be formulated as: To create an outstanding image of Gateshead.

The Creative Response – the Vision

Contextual Constraints

- Establish relationships with client team and socially affected groups
- Selection of team with necessary professional competence, knowledge of natural laws
- Surveys of context – physical, climatic, topographical, materials, energy
- Legal requirements – planning permissions, legislation.
- Funding and value for investment, competition
- Programme requirements
- Health and safety issues
- **Summary of contextual items**

Anthony Gormley was selected from a short list of artists and visited the site at the head of the Team Valley overlooking a long sweep of the A1 motorway.

The commission was confirmed in 1994, but there was considerable local opposition. In 1996 Northern Arts, the regional arts coordinator, staged a year-long series of exhibitions and events throughout the north. As part of this, Anthony Gormley staged a major exhibition of small terracotta figures. This was very well attended and led to a more positive view of the *Angel*.

Effective Creativity

- Interdisciplinary discussions
- "Brain storming" sessions
- Preparation of alternative imaginative schemes
- Programme for delivery
- **Selection of a preferred solution and its clear definition**

The Council were able to move ahead with their plans. The sculptor conceived the *Angel* as using heavy industrial and shipbuilding techniques to create an industrial image of a spiritual subject. Aeronautics and anatomy combine to form an exoskeleton of ribs and diaphragms and an inner body of plate modelled on the sculptor's own body.

To proceed with the development, Arup's Newcastle office were commissioned to advise on the structural design. They were able to ensure that the large and vulnerable structure would remain standing under extreme conditions of gale force winds on the top of the hill. The production of a viable structure required close collaboration between sculptor and engineer and a clear understanding of the fabrication and construction problems. State-of-the-art technology was needed to produce the final construction scheme.

A corten steel was adopted which does not need painting but is protected by a rusty patina that forms during the first few years.

A 3D model of the body was developed, and instructions from this were passed directly to the cutting machine which produced the pieces of plate from which the *Angel* was made.

Figure 44. Completed statue of the Angel of the North, showing foundations

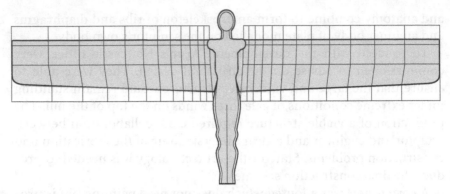

Figure 45. Dimensioned elevation

The reclaimed colliery pit head site has up to 15 m of fill over rockhead. Two coal seams were identified beneath the site. These workings were grouted up and a foundation of eight 750-mm bored piles end bearing in the rock were driven. The pile cap of 12×8×1.5 m on these piles ensured that all piles remain in compression even under extreme wind loading.

A budget for the sculpture was estimated, but because of the complexity of the project, potential fabricators were invited to advise on the feasibility and likely costs. Local contractors were selected. The foundation costs were also significant, having to take into account existing old mine workings, requiring deep foundations. These assembled costs then enabled the Council to start fund-raising.

The scheme was confirmed as the major sculpture – 20 m high and 54 m wide – and the selection of suitable fabricators and constructors could begin (Figure 44).

 # Delivery

- Agreement and documentation of preferred solution – continuous reviews of delivery against aims – the design audit
- Definition of management team and process of delivery
- Selection of key designers, of contractors, sub-contractors, and specialists
- Definition of their roles
- Reference to interested third parties

Figure 46. Completed statue

- Agreement on financing – budget, cash flow, cost control procedures
- Agreement on delivery programme
- Agree contract procedures
- Assembly of project/product resources – skilled crafts, materials, equipment
- Agree quality management procedures – design, manufacture, construct
- Health and safety considerations
- Legal requirements and programme
- **Hand over to clients**

The development of the fabrication and erection techniques required very close collaboration between the sculptor, the engineers and the contractors. This enabled a design to be formulated that could be fabricated and assembled by local contractors. The sculptor maintained a close interest in the fabrication processes, encouraging the craftsmen and ensuring a high standard of finish. Sophisticated computer modelling techniques were adopted to confirm the interaction of the several components, and the viability of the proposed erection techniques. The three major prefabricated components were delivered to site by road, arousing a great deal of local interest, which continued during the erection and assembly processes. These were erected using 500 tonne and 300 tonne cranes, watched by assembled TV crews, journalists, photographers and local spectators!

 # Performance in Practice

- Reassessment of design brief
- Possible design development
- Social impact assessment
- Client/user satisfaction
- Economic performance
- Maintenance/operational factors
- Experience to incorporate in future projects

Angel of the North
vital statistics
Steelwork
· Height: 20 m
· Wingspan: 54 m
· Average wing depth: 5.7 m
· Weight: 208 T
· Ankle cross-section: 780 mm x 1.4 m
· 3000 pieces of steel assembled
· 136 bolts each 48 mm diameter attach wings to body
· 22 000 man-hours on fabrication
· 10 km of welding

Design
· 70 T horizontal wind force to be resisted
· 450 T force in wing diaphragms
· 1200 T force in ankle ribs
· 50 T force in each 50 mm bolt
· 2500 man-hours on engineering design and drawing

Foundations
· 5000 m³ soil excavated and replaced to reform mound
· 100 T grout pumped into mineworkings up to 33 m below ground
· 700 T concrete and 32 T reinforcing steel in foundations to 20 m below ground
· 52 bolts each 50 mm diameter and 3 m long hold the Angel upright in wind

Figure 47. Vital statistics

The interest in the project has continued and has been a significant factor in attracting lottery funding for a major new art gallery and concert hall to be built on the banks of the Tyne, together with a footbridge to link the complex to Newcastle Quayside. The project has confirmed the importance of close liaison between the arts and sciences. The status of Gateshead has improved significantly as both a thriving industrial town and a community with strong artistic awareness.

5 | Project Design Audit

This chapter demonstrates the design process and suggests how it can be audited during the development of a solution to meet a need. The framework of the three basic stages in the process has been described in detail in Chapter 2. Since the process of developing a solution to a need can take several years, it is of value to have a checking process which audits the work done to ensure that the key items are all covered, and that the original intention is maintained.

 ## Values

The design process cannot be detached from ethical and moral aspects of decision-making. The design process set out in the Appendix and developed in Chapter 2 must, however, be complemented by basic value judgements if it is to respect and improve the quality of life in human societies, and take care of the world in which we live.

At the heart of this idea of respect is the recognition of common humanity; all affected are considered as having equal value. This may be summed up in terms of the "Golden Rule", a concept common to many different ethical systems and cultures including:

The Christian version "Treat others as you would like them to treat you" (Luke 6, 31); "Love your neighbour as yourself" (Matthew 22, 39)

Hindu version "Let not any man do unto another any act that he wisheth not done to himself by others, knowing it to be painful to himself" (Mahabharata, Shanti Parva)

Confucian version "Do not do to others what you would not want them to do to you" (Analects, Book xii, 2)

Buddhist version "Hurt not others with that which pains yourself" (Udanavarga, v.18)

Jewish version "What is hateful to yourself do not do to your fellow man. This is the whole Torah" (Babylonian Talmud, Shabbath 31a)

Muslim version "No man is a true believer unless he desires for his brother that which he desires for himself" (Hadith Muslim, imam 71-2)

In these versions of the Golden Rule, there is a balance of concern for the self and the other which is in essence unconditional. It does not depend upon any quality or group or type of person. In general, the balance of the Golden Rule reminds us that the needs and value of the individual have to be taken into account as well as those of others and of the creation as a whole.

 # Professions

The nature and role of the professions is discussed in Chapter 1. Engineering decision-making is clearly concerned with international and global issues, with varying needs and traditions. As engineering decision-makers we must audit our activities to ensure that the overall needs of humanity are recognised and the global consequences are considered.

 # A Detailed Case Study

An example of the design issues that arise during the several stages in the procurement of a major project, in this case the Falkland Islands Airport, has been selected and is set out below. Suggestions are made as to how an effective audit of the process in detail can be organised following the processes of design as discussed in Chapter 2. The text comments upon each of the three stages of need, vision and delivery, showing how decisions are made relative to the checklists for each stage. The same checklist can be used to monitor and audit progress regularly throughout the project development.

The several case studies given in Chapter 4 are also useful as examples of the strategic design process. This audit shows how the review of the process to maintain the established aim can be carried out from time to time during the process of project development.

Formulation of Need

- The cultural societal and physical ethos in which the project is to be carried out
- The history of the project, how did the perception of need arise?
- The decision-making structure – social, client, design management
- The human resources available – professional skills, research, available craft skills
- The physical, technical and economic resources available
- Sustainability requirements, energy sources, environmental impact
- Relevant research and development
- **The clear formulation of the need**

The Project Ethos and History

As with all major public sector projects, several ministries were concerned with this project. In this case they included the Foreign Office, the Ministry of Defence, the Department of the Environment via the Property Services Agency, and the Treasury. Professional engineers and others involved were directly employed, some by the government departments and others by the appointed consultants and contractors. All the basic design work was done in the UK, including the site location of the airstrip and the basic layout. The appointed consultants then developed the site facilities, airport and accommodation buildings and communication facilities. Multi-disciplinary design sessions were organised to ensure the development of an integrated scheme.

The aims of the client departments were to achieve an acceptable operational airport, for both public and defence use, to demonstrate an openness in the international implications of the project, to reassure the islanders that their affairs were being considered and that the quality of their lives would not be reduced by the provision of these extended transport facilities, and to obtain value for money and full accountability.

It was a condition of the contract that only UK consultants and contractors would be engaged, and that materials and supplies would all come from the UK. Some aspects of the project were subject to security restrictions, requiring discretion.

The primary aim in constructing the airfield at Mount Pleasant in the Falkland Islands was to secure the security of the Islands, following the Argentinean invasion, by providing for a rapid reinforcement of the garrison in the event of a crisis, enabling the size of the resident garrison to be reduced. The airfield would also provide direct access to the Antarctic, and if need be to Australasia without having to land on any non-British airfield. Alternative sites were carefully considered before selecting the site at Mount Pleasant.

Decision-making Structure – Skills Available

Team members were selected to ensure that there was the professional competence to perform all the work undertaken and maintained and improved by continually updating and extending the understanding of technical and professional developments. Clearly, on such a high speed project, the need for the professionals to have the confidence of the client and to respect each other's integrity and expertise was paramount. It was essentially a major team project, bringing together as partners organisations that had not necessarily previously collaborated, but all of whom knew and trusted the quality of the work of the others. The teams assembled of client representatives, consultants, contractors and suppliers were all very experienced, and in many cases had collaborated previously on other public sector projects.

Factors considered at this stage included speed of construction, cost control, improvement of trade, disturbance to the peaceful environment, effect on farming, changes in social structure, profitability, logistics, reduction of risk (commercial), enhancement of reputation, work prospects, higher rates of pay, living conditions, the effects on indigenous flora and fauna.

Those factors which were likely to impact on the project, including all the factors arising during the lifetime of the project, needed to be discussed by all parties when formulating the need so that convergence into an agreed overall brief could be prepared. At this stage we needed to note the differing "missions" of the several parties – Defence Department, Treasury, security, confidentiality, commerce, local community, contractors, suppliers, consultants, workers, ecologists – were all considered. Because speed was of the essence in this project, the client sought out practitioners with proven prior experience and competence in the type of construction work involved.

Only concrete aggregates were obtained locally, and these required the development of new borrow pits and quarries, with significant environmental constraints.

Physical Factors

The airfield was sited some 55 km from Port Stanley, requiring the construction of a new road to link it to the port facilities. This isolation helped to protect the small community of Port Stanley from the major impact of the very large work force.

The particularly unusual features of the project were the remoteness of the site, with the need to simplify the complicated logistical problems, to allow for the staffing problems of travel, isolation, local care, policing, *etc.*, and the need to make a minimal impact upon the rather fragile local ecosystem, both human and ecological.

Sustainability

The consequences of such a project are very far reaching and need to be fully considered in its planning and execution. The impact of the construction work was considerable. The quarries for example were a much more significant single development than had ever taken place before in the Islands. Steps were taken to ameliorate the worst effects of the development upon the environment, and to create a self-contained way of life within the construction and operational communities.

It is not unlikely that in the future the development of some of the natural resources of the South Atlantic may become financially viable. In this event the availability of the improved transport facility will be very valuable, not only to the Falklanders, but to the world community at large. It will however also encourage, perhaps adversely, the exploitation of the resources, and care will need to be exercised to ensure that this is not overdone.

The audit reviewed the monitoring of the development of the agreed formulation at intervals to ensure that the original aims were being met. The formulation is given below as the conclusion of the need section of the design process.

Formulation of need: To provide effective airport facilities to improve the security and develop the economy of the Islands.

The Creative Response – the Vision

Contextual Constraints

- Establish relationships with client team and socially affected groups
- Selection of team with necessary professional competence, knowledge of natural laws
- Surveys of context – physical, climatic, topographical, materials, energy
- Legal requirements – planning permissions, legislation.
- Funding and value for investment, competition
- Programme requirements
- Health and safety issues
- **Summary of contextual items**

Relationships Between All Parties, Legal Requirements
The airport project had a profound effect upon the social life of the Falklands community. The total population of the Islands was just under 2000 at the time of the conflict with Argentina, of whom about 900 lived in the largest town of Port Stanley. Their chosen and preferred way of life centred around sheep farming, with a few additional service jobs and a little fishing. The way of life was similar to that of some of the islands off the coast of Scotland. Leisure activities were simple and communal, involving the whole family, as illustrated by the regular Saturday evening dances and social events which were attended by all generations.

Government and justice were easily administered. The natural resources of the Island are limited, little cultivation having taken place. The community are dependent upon imports for most of the commodities needed for their daily lives. The general level of incomes was low, but adequate for the needs of the Islanders. It was to this sheltered community that more than 3000 workers came to build the airport, affecting all aspects of the community's way of life. It was necessary to integrate matters of health and safety with the island community. Hospital facilities were limited, and the health care aspects of the contract provisions were very important.

It was important to develop and maintain good relationships with neighbouring countries, particularly with Brazil and Argentina. The change of government in Argentina made this possible before the airport was completed.

Selection of Team, Financial Issues

Because of the unusual nature of the project, attention was focused on the major issues, and they were fully discussed before the start of the contract, which made the definition of procedures and the management of the site operations a recognised part of the regular progress reviews throughout the project. It was particularly important to make all financial arrangements transparent for reporting to the UK Treasury.

In considering the implications of the design stage decisions in the case study of the Falklands project, many unusual factors were taken into account. The total context of the project, because of the very small economic "eco-system" of the Islands, had a considerable effect upon the community. The commercial and management aspects of the project were very important. The key decision was to insulate the local economy from the project in so far as this was possible. In the consideration of alternative sites, attention was given to ensuring that where possible the decisions could be reversed, and that sustainability was an essential feature of all aspects of the work. Health and safety were also crucial issues.

The Falkland Islands are renowned for their unusual variety of wildlife and as breeding grounds for some unusual and significant birds and marine mammals. An independent, thorough environmental impact analysis of the project was commissioned and carried out in parallel with the design process. This enabled certain important breeding grounds to be identified. These required that areas be identified for the extraction of aggregates and for the opening of a quarry to provide materials for concrete and for the surfacing of pavements.

Removing the airport from the immediate vicinity of any settlement has had the effect of reducing the impact, but some permanent adjustment will be inevitable.

The wealth of the whole community has been increased, but will need to be carefully monitored to ensure that the economy remains in balance. The community is in much closer touch with the rest of the world than had previously been the case.

Surveys of Context

Most of the land is undeveloped moorland peat bogs, with outcrops of rock and scree. There are a considerable number of small and shallow lakes, but few major streams. There is a very long and indented coast-line. The native flora and fauna are limited in variety, but include some rare and interesting species of South Atlantic life forms. In establishing the basic design and the agreed construction procedures, the contextual constraints were defined and referred to regularly as part of the audit procedure.

Effective Creativity

- Interdisciplinary discussions
- "Brain storming" sessions
- Preparation of alternative imaginative schemes
- Programme for delivery
- **Selection of a preferred solution and its clear definition**

It was necessary to locate an adequate water supply for the project using local streams; recognising that the operational demand would be less than the construction demand, the provision had therefore to be easily reversible on completion of the project.

All these special factors were considered in design team sessions. The avoidance of conflict depends upon good communication, upon the agreement of common aims, and upon the trust between the several parties involved in the project.

The completed designs were officially "signed-off" by the design team and client, and described in a definitive specification and well-presented schedules and drawings. A reference copy was carefully filed, so that any subsequent changes that may prove unavoidable could be adequately cross-referenced to the original proposals. This was an essential component of the auditing process.

Careful consideration was given in the design of the facilities to ensure that buildings essential to construction, for example, staff accommodation, harbour facilities, recreational facilities, chapels, clubs, *etc.*, could either be converted into permanent facilities or removed on completion. The time scale for the project was short. For example, a period of only about 10 weeks was available for the production of contract documents for approximately £200 million of work. The usual airport facilities for long haul passengers were included in the development. Good communication facilities were essential; they worked so well in fact that we found it easier to send drawings to the Falklands than between our UK offices!

Delivery

- Agreement and documentation of preferred solution – continuous reviews of delivery against aims – the design audit
- Definition of management team and process of delivery
- Selection of key designers, of contractors, sub-contractors, and specialists
- Definition of their roles
- Reference to interested third parties
- Agreement on financing – budget, cash flow, cost control procedures
- Agreement on delivery programme
- Agree contract procedures
- Assembly of project/product resources – skilled crafts, materials, equipment
- Agree quality management procedures – design, manufacture, construct
- Health and safety considerations
- Legal requirements and programme
- **Hand over to clients**

The Contractors – Definition and Selection of Effective Construction Team

Working procedures between the project team, particularly the on-site management, were set up to ensure that a full understanding between the construction team and the local interests was established and maintained. These included the farming and fishing interests as well as the local security services (established garrisons and the police force). The impact upon local amenities, recreation facilities, employment and resources was also carefully considered. Regular meetings took place between the project management team and the local representatives.

All the contractors involved in the project were very experienced in operating in remote locations, and were known to have the resources needed to carry out the work. They were also experienced in public sector work, and the procedures needed to reduce misunderstandings and potential conflicts to a minimum, and to cope with contingencies where necessary.

Senior level communication was established with the local government and community officers to ensure that external relationships were good. Where possible sharing of facilities was arranged, enhancing the opportunities for the improvement of community activities. Particular attention had to be given to the provision of accommodation and recreational facilities for the site staff, and morale was maintained at a high level throughout the project.

Pre-tender briefings and site visits helped to ensure that there was a full understanding of all restrictions on material and labour resources, and of the health and safety requirements for this unusual project.

This careful preparation helped considerably in securing full collaboration between all the parties as members of a very inter-dependent work force, provided with all accommodation and recreational facilities. It was particularly important to establish effective communication links with interested third parties.

Because of the special circumstances, the need for close liaison between members of the project team and the local community was recognised. Considerable authority was delegated to the site staff, both supervisory and contractual for both external and internal relationships.

Many of the construction elements had to be fabricated in the UK and shipped out, to help to limit the size of the resident labour force. In spite of this close collaboration, it was essential, in such a high profile project, that full public accountability be maintained to ensure the good husbandry of public resources and demonstrable value for money.

Assembly of Resources – Employees, Materials, Equipment

The needs of employees of the construction team were carefully considered. No local labour was to be employed, because of possible serious impact on the island economy. This applied to all employees: the

PSA, consultants, contractors, sub-contractors, and suppliers. The responsibilities of the management team for the well-being and happiness of the work force become of particular importance. Where necessary, training facilities were provided to relate the work force to the special needs of the project. Operatives had no possibility of leaving the project to have a life of their own, as would be the case on a home-based project. Tour periods were of about 6 months' duration, almost entirely on a bachelor basis.

Job satisfaction, the maintenance of career developments for the professionals, the balance of the rewards by the differing employers, and the provision and operation of common residential and recreational facilities and of health and safety facilities were all important factors in ensuring the smooth running of the project. An effective security/police force formed part of the community structure.

Since for both political and practical reasons all materials and labour were brought from the UK, a separate handling facility was needed, and the excellent harbour at Port Stanley was not available. A temporary harbour was created by "permanently" mooring one of the first supply vessels, with adequate handling equipment, at Mare Harbour, about 7 km from the selected airfield site. This acted as a jetty, and all subsequent cargoes were unloaded across its deck, which was connected by a bridge to the shore. This was later replaced by a permanent berthing facility.

For similar reasons there was a need to ensure that subsequent maintenance of the facilities was reduced to a minimum and capable of being undertaken by a local team of operatives, without access to sophisticated equipment or the ability to purchase any necessary materials from the local community. Prefabricated timber buildings were adopted for the residential buildings, and simple steel-framed buildings for the workshops, power house, aircraft hangers, *etc.* When the logistics, programme or financial constraints on projects are particularly difficult, then the use of tried and tested construction techniques and materials is usually preferable. Quality control was established on site and operated as an independent unit, there being no separate facility available as would be the case in most large UK-based projects.

The close involvement of the site staff in the earlier stages of decision-making, and the close and continuous communication between site and home-based teams (clients, suppliers, consultants and others) ensured that decisions could be made which did not conflict with the strategic criteria of the client body in the UK.

Construction workers frequently take a great pride in their work, and it was observed on the arrival of the first aircraft on the airstrip that many of the construction workers lining the runway to welcome the plane were weeping with delight at their achievement!

Appendix – The Case Study Format

The factors to be considered at each stage are set out briefly below.

 ## Formulation of Need

- The cultural societal and physical ethos in which the project is to be carried out
- The history of the project, how did the perception of need arise?
- The decision-making structure – social, client, design management
- The human resources available – professional skills, research, available craft skills
- The physical, technical and economic resources available
- Sustainability requirements, energy sources, environmental impact
- Relevant research and development
- **The clear formulation of the need**

 ## The Creative Response – the Vision

Contextual Constraints

- Establish relationships with client team and socially affected groups
- Selection of team with necessary professional competence, knowledge of natural laws

- Surveys of context – physical, climatic, topographical, materials, energy
- Legal requirements – planning permissions, legislation.
- Funding and value for investment, competition
- Programme requirements
- Health and safety issues
- **Summary of contextual items**

Effective Creativity

- Interdisciplinary discussions
- "Brain storming" sessions
- Preparation of alternative imaginative schemes
- Programme for delivery
- **Selection of a preferred solution and its clear definition**

 # Delivery

- Agreement and documentation of preferred solution – continuous reviews of delivery against aims – the design audit
- Definition of management team and process of delivery
- Selection of key designers, of contractors, sub-contractors, and specialists
- Definition of their roles
- Reference to interested third parties
- Agreement on financing – budget, cash flow, cost control procedures
- Agreement on delivery programme
- Agree contract procedures
- Assembly of project/product resources – skilled crafts, materials, equipment
- Agree quality management procedures – design, manufacture, construct
- Health and safety considerations

- Legal requirements and programme
- **Hand over to clients**

Performance in Practice

- Reassessment of design brief
- Possible design development
- Social impact assessment
- Client/user satisfaction
- Economic performance
- Maintenance/operational factors
- **Experience to incorporate in future projects**

Index